Conservation of Saline Wetland Ecosystems

Laxmi Kant Sharma • Rajashree Naik

Conservation of Saline Wetland Ecosystems

An Initiative Towards UN Decade of Ecological Restoration

Springer

Laxmi Kant Sharma
Central University of Rajasthan
Rajasthan, India

Rajashree Naik
VIT-AP University
Amaravati, Andhra Pradesh, India

ISBN 978-981-97-5068-9 ISBN 978-981-97-5069-6 (eBook)
https://doi.org/10.1007/978-981-97-5069-6

This Springer imprint is published by the registered company Springer Nature Singapore Pte Ltd.

The registered company address is: 152 Beach Road, #21-01/04 Gateway East, Singapore 189721, Singapore

This book is dedicated to our beloved parents.

Foreword 1

I am pleased to introduce this important work titled *Conservation of Saline Wetland Ecosystems*, authored by Professor Dr. Laxmi Kant Sharma from the Central University of Rajasthan and Dr. Rajashree Naik from VIT-AP University. This comprehensive exploration sheds light on the often-underestimated saline wetlands, emphasizing their crucial role in supporting the maintenance of our planet's ecological balance.

As we confront increasing environmental challenges, understanding and preserving these unique ecosystems are of paramount importance. This book is poised to constitute a significant contribution to the field of environmental conservation, offering both in-depth analysis and practical insights into the preservation of saline wetlands.

Professor Sharma, a distinguished figure in environmental science, brings decades of expertise and advocacy for sustainable ecosystems to this collaborative effort. The partnership with Springer Nature further underscores the global significance of this work within the scientific community.

I believe that this publication's accessibility and its very useful content will render it an invaluable asset to the scientific community, leaving a lasting impact on discussions surrounding environmental conservation.

Youssef Nassef
Adaptation Division
UN Climate Change (UNFCCC) Secretariat,
Bonn, Germany

Foreword 2

In the realm of environmental preservation, the significance of our wetland ecosystems cannot be overstated. They serve as vital reservoirs of biodiversity, playing a crucial role in maintaining the delicate balance of our planet's ecological health. Among these ecosystems, saline wetlands stand as unique and often overlooked treasures, harboring a diverse array of life adapted to their challenging saline environments.

This book *Conservation of Saline Wetland Ecosystems* by Professor Laxmi Kant Sharma from Central University of Rajasthan and coauthored by Dr. Rajashree Naik from VIT-AP University is a timely and indispensable contribution to our understanding and preservation of these delicate habitats. Through meticulous research, profound insights, and a deep passion for environmental conservation, Professor Sharma and Dr. Naik illuminate the complexities and nuances of these remarkable ecosystems.

This book serves as a guiding beacon for policymakers, conservationists, researchers, and all those dedicated to the protection of our planet's natural heritage. The authors' expertise, honed through years of dedicated study and practical experience, shines through in these pages, offering a roadmap for sustainable management and preservation strategies.

As we navigate the challenges posed by rapid environmental changes, the wisdom encapsulated in this book becomes increasingly pertinent. It is a call to action, urging us to recognize the intrinsic value of saline wetlands and to undertake collective efforts for their conservation and sustainable utilization.

It is a great pleasure for me to present the invaluable contribution of Prof. Laxmi Kant Sharma and Dr. Rajashree Naik towards the conservation of wetlands, and I am confident that this seminal work will inspire and guide generations in safeguarding these invaluable saline wetland ecosystems for the benefit of all life on our planet.

Anand Bhalerao

Vice Chancellor
Central University of Rajasthan,
Kishangarh, Ajmer, Rajasthan, India

Foreword 3

In the following pages, you are about to embark on a journey of discovery and dedication—a journey into the heart of one of our most precious and imperiled ecosystems, wetlands. *Conservation of Saline Wetland Ecosystems* is a testament to the deep connection we share with these vital landscapes and an exploration of the profound responsibility we bear in protecting and preserving them.

Saline wetlands, often overlooked and underestimated, are the lifeblood of our planet. They are the vibrant crossroads where land and saline water meet, giving birth to unique ecosystems teeming with biodiversity. This book, meticulously compiled by the authors, offers a window into the critical role saline wetlands play in the tapestry of our world.

As you navigate through these pages, you will encounter stories that illustrate the urgency and beauty of saline wetland conservation. Each chapter unveils a new layer of understanding, emphasizing the significance of saline wetlands and the tireless efforts of individuals committed to their safeguarding. Saline wetlands offer us more than just their intrinsic beauty; they provide us with essential services. These precious environments play a pivotal role in mitigating the impacts of climate change and preserving the delicate balance of our world. This book seeks to inspire you, as it has inspired me, to become a guardian of saline wetlands. By joining the ranks of those dedicated to their preservation, we ensure that these irreplaceable ecosystems will continue to thrive for generations to come.

Thank you for embarking on this enlightening journey through the pages of *Conservation of Saline Wetland Ecosystems*. Your commitment to the conservation of wetlands is a testament to your dedication to the well-being of our planet and the legacy we leave for future generations.

VIT-AP University
Amaravati, Andhra Pradesh, India

S. V. Kota Reddy

Preface

The first author, Professor (Dr.) Laxmi Kant Sharma native place is near to the Thar Desert of India (*Renwal, Jaipur, Rajasthan*). It has a very unique water body, named as Sambhar Lake. It is a seasonal inland saline wetland which attracts lakhs of flamingoes and other migratory birds every winter season. From the very childhood, the author and his friends were fond of bird watching in the rich pink landscape of the lake. By the time he graduated, he observed the number of visiting birds to the lake is reducing along with their diversity. Then, he went for his higher studies and profession to other parts of the country. In 2015, when he returned to his home state as a Professor to the Department of Environmental Science, Central University of Rajasthan, Ajmer, he was shocked to see the shrunken lake. He felt so disheartened to know that lake is not able to welcome its lakhs of migratory birds.

Being a responsible and dedicated environmentalist, the author decided to revive his childhood love for lake. Though his core research interest has been for forest ecosystems, he extended himself for wetland ecosystem. Till then, he has never looked back. He set his big goal of conservation and restoration of lake but started with incrementally small steps. He started to train the postgraduate students of environmental science, through their dissertation work. After 3 years of baseline research, in 2018, he started to train Ph.D. students for this noble goal.

The second author, Rajashree Naik, is the first Ph.D. scholar, who intensively worked for the Sambhar Lake using integrated approach of remote sensing and geospatial technology along with ground observation and local people's perspective toward this lake. The study revealed that the lake has been shrinking approximately 4.23% every decade since 1972 and it might be completely lost by 2059 if not taken care off. During her doctoral journey, both the authors realized that not only Sambhar Lake but also other saline lakes are undergoing the similar deteriorating situations. The authors started to draw attentions of the academicians, scientists, and governments at local, national, and global level through research papers in the international journals such as PLoS One, Wetland Ecology and Management, Journal of Oceanology and Limnology, news articles in The *Times of India* and *The Hindu*, encyclopedia articles in Wiley publication house, and interviews at television channels such as *Zee News*. The dedication was enhanced when the authors received four international awards and grants from the *Society of Wetland Scientists*, *British Ecological Society*, *Ecological Society of America*, and *International Society of Limnology*. The motivation further increased when the Ph.D. work of the second author was acknowledged by Ramsar Convention, on it 50th anniversary, in 2021, and he was invited as keynote speakers in reputed platforms such as *Society of Wetland Scientists* (SWS), *Youth Engaged in Wetlands* (YEW), and *Young Ecosystem Services Specialists* (YESS).

While the authors successfully highlighted these unique yet often overlooked ecosystems, they recognized the urgent need to educate the public comprehensively about their characteristics, ecosystem services, the threats they face, and the application of advanced technology and artificial intelligence for their restoration and conservation for future generations. Since it is widely accepted that "Pen is mightier than sword," the authors stepped forward toward giving a physical shape to their love, knowledge, and experience for the saline wetlands, which landed in the form of this wonderful book.

This book contains basics to advanced-level information about saline wetland ecosystems, and no pre-requisite knowledge is required to read this book. This book is divided into four units based on the theme, i.e., ecology or technology. The first unit provides a glimpse of the ecological importance and threats about these lakes, the second unit deals with the past, present, and future technological usage for these wetlands, the third unit showcases Sambhar Lake (childhood love of Laxmi Kant Sharma) as a case study, and the last unit deals with the restoration approaches that can be undertaken during the UN Decade on Ecosystem Restoration for these neglected ecosystems.

Since, this book contains a lot of interesting infographics and tables, the authors are very much confident that it will be beneficial to the large range of audience from students, researchers, academicians, scientists, environmentalists, nature activists, NGO personnel, and general public, who may or may not be related to the wetland ecosystems. This book will not only provide relevant information but also motivate the readers to actually work on the ground toward their sustainable restoration, management, and conservation for the present and future generation.

Ajmer, Rajasthan, India Laxmi Kant Sharma
Amaravati, Andhra Pradesh, India Rajashree Naik

Acknowledgements

We extend our heartfelt gratitude to the following individuals and organizations whose invaluable contributions have played a pivotal role in the creation of this book, "Saline Wetland Ecosystems as Alternative Options for Future Generation: An Initiative Toward UN Decade on Ecosystem Restoration."

We express our sincere appreciation to Professor Anand Bhalerao, Vice-Chancellor of the Central University of Rajasthan, for his unwavering support and encouragement throughout the research and publication process. His commitment to academic excellence has been a guiding force. Our gratitude extends to Dr. S.V. Kota Reddy, Vice-Chancellor of Vellore Institute of Technology University, Andhra Pradesh (VIT-AP), for his support and collaboration. His commitment to fostering research initiatives has been instrumental in the success of this project. We would like to express our gratitude to our colleagues whose collaboration and insightful discussions have greatly enriched the content of this book. Their diverse perspectives and expertise have been invaluable.

We acknowledge the Ramsar Convention on Wetlands for their dedication to the conservation and sustainable use of wetlands worldwide. Their commitment to promoting ecological restoration aligns seamlessly with the goals of this book. Our in-depth gratitude goes to Wetlands International for the continuous efforts in wetland conservation and restoration. Their expertise and partnership have enriched the content of this book and contributed significantly to its quality.

Our sincere thanks to Springer Nature Publication House for their professionalism, expertise, and commitment to academic publishing. Their support in bringing this work to fruition is deeply appreciated. Finally, to the parents and family members of the authors, we extend our deepest appreciation for their unwavering support, understanding, and encouragement. Their belief in our capabilities has been a constant source of motivation.

The collaborative efforts of these individuals and organizations have been indispensable in shaping this book. Their commitment to the advancement of knowledge in the field of saline wetland ecosystems is a testament to the collective endeavor toward a sustainable and ecologically restored future.

Contents

Abbreviations

3D	Three dimensions
AI	Artificial intelligence
ALOS	Advanced Land Observing Satellite
AMS	Accelerator Mass Spectrometry
AVIRIS-NG	Airborne Visible InfraRed Imaging Spectrometer-Next Generation
CASI	Compact Airborne Spectrographic Imager
CHM	Canopy Height Model
CNES	Centre National D'Etudes Spatiales
CNN	Convolutional Neural Network
CV	Computer Vision
DEM	Digital Elevation Model
DInSAR	Differential SAR Synthetic Aperture Radar Interferometry
DO	Dissolved Oxygen
DOM	Dissolved Organic Matter
DSM	Digital Surface Model
EMI	Electro-Mmagnetic Induction
EnMAP	Environmental Mapping and Analysis Program
ESA	European Space Agency
EVI	Enhanced Vegetation Index
EWS	Early Warning Systems
FAO	Food and Agriculture Organization
GCP	Ground Control Points
GHG	Greenhouse Gases
GIS	Geographic Information Systems
GNSS	Global Navigation Satellite Systems
GPR	Ground-Penetrating Radar
HABs	Harmful Algal Blooms
IBA	Important Bird Area
ICESAT	Ice, Cloud, and land Elevation Satellite
InSAR	Interferometric Synthetic Aperture Radar
IoT	Internet of Things
ISRO	Indian Space Research Organisation
KOMPSAT	Korean Multi-Purpose Satellite
LAI	Leaf Area Index

LiDAR	Light detection and ranging
LULC	Land Use Land Cover
MODIS	Moderate Resolution Imaging Spectroradiometer
MoEFCC	Ministry of Environment, Forest and Climate Change
NASA	National Aeronautics and Space Administration
NDMI	Normalized Difference Moisture Index
NDVI	Normalized Difference Vegetation Index
NDWI	Normalized Difference Water Index
NGO	Nongovernmental organizations
NIR	Near-Infrared
NLP	Natural Language Processing
ORP	Oxidation–Reduction Potential
Ppt	Parts Per Trillion
PRISMA	PRecursore IperSpettrale della Missione Applicativa
PSI	Persistent Scatterer Interferometry
RF	Random Forest
RNN	Recurrent Neural Network
RS	Remote Sensing
SAR	Synthetic Aperture Radar
SDGs	Sustainable Development Goals
SDM	Species Distribution Modeling
SPOT	Satellite Pour l'Observation de la Terre
SVM	Support Vector Machine
TSS	Total Suspended Solids
UAV	Unmanned Aerial Vehicles
UNEP	United Nations Environment Programme
USGS	United States Geological Survey
WWF	World Wildlife Fund

List of Figures

List of Tables

Part I

Study of Wetlands and Its Types

Wetland Ecosystems

1

Abstract

This chapter of the book, titled "Wetland Ecosystems", initiates a thorough investigation into the profound significance of these often overlooked yet vital natural habitats. We delve into the multifaceted aspects of wetlands, unveiling their ecological treasures and paramount importance to both the environment and humanity. Divided into several sections, each offers a distinctive lens through which to view wetlands. The importance of wetland ecosystems is thoroughly examined, showcasing their diverse ecological services ranging from wildlife habitat provision to water purification and flood control. Amid the backdrop of the climate crisis, we explore the indispensable role wetlands play in mitigating its impacts, serving as crucial carbon sinks and resilient ecosystems. Furthermore, readers are introduced to the key organisations dedicated to wetland conservation and the pivotal Ramsar Convention, which provides a framework for global cooperation in wetland preservation. The global distribution of wetlands and Ramsar Sites is meticulously explored, alongside a detailed classification based on hydrology, vegetation, and geography. Highlighting the manifold ecosystem services offered by wetlands underscores the imperative of their preservation, despite the myriad threats they face, including habitat destruction, pollution, and climate change. This chapter advocates for concerted efforts to safeguard these invaluable ecosystems for future generations.

Keywords

Climate Crisis · Ecosystem Services · Global Wetland Distribution · Ramsar Convention · Wetland Ecosystem

L. K. Sharma, R. Naik, *Conservation of Saline Wetland Ecosystems*,
https://doi.org/10.1007/978-981-97-5069-6_1

Water-saturated soils are a defining feature of wetland ecosystems and contribute to their unique diversity and constant change. They appear in many different forms, such as marshes, swamps, bogs, and floodplains, and can be found anywhere from the coasts to the interiors of the continents. Unique features distinguish these ecosystems from their terrestrial and aquatic counterparts. One of the defining features of the area is the prevalence of hydric soils, which are soils that are entirely saturated with water for a significant portion of the year. Waterlogged conditions produce an oxygen-depleted atmosphere, which in turn influences the kind of plants and animals that may survive there. Wetlands are home to a wide variety of plants. There is a broad variety of hydrophytic (water-loving) plant species present, all of which have adapted to thrive in anaerobic conditions. Water lilies, cattails, reeds, and sedges are all examples of aquatic vegetation. Adaptations that wetland plants have developed over time allow them to survive in wet, waterlogged soils and nutrient-rich environments. Furthermore, wetlands serve as vital water storage zones. Water levels fluctuate throughout the year because they are either fed by surface water runoff or directly connected to groundwater sources. They often obtain their water supply from surface water discharge. The presence of water not only affects plant and animal life, but also provides habitats critical to supporting a wide variety of species and ecosystems.

There is a wide range of biological functions performed by wetland ecosystems, many of which are important for global health. One of its primary roles is to act as a natural filter, cleansing the water as it passes through them. Wetlands are comparable to sponges in that they remove sediments, nutrients, and toxins from the water by soaking them up and holding on to them. Downstream rivers, lakes, and coastlines benefit from this filtration process because it helps keep them clean. During dry periods, wetlands gently release their stored water, which serves to regulate the flow of rivers and streams in the ecosystems that border the wetland. Wetlands are also significant because they serve as homes for a wide diversity of plant and animal species. The presence of water, nutrient-rich soils, and particular flora allow for the survival of a wide range of rare and, in many cases, threatened species. Many different kinds of animals, both vertebrate and invertebrate, live in wetlands. These ecosystems are crucial for migratory birds as breeding, nesting, and feeding areas. Furthermore, these habitats are critical resting spots for migrating birds on their lengthy journeys. The incredible array of flora and fauna that may be found in the wetland has attracted a lot of interest.

They aid in keeping the delicate ecological balance by serving as safe havens and breeding grounds for several species. Wetlands provide a variety of habitats, including open water, marshes, and wet meadows, that are suited for a wide range of plant and animal species. The plant life that flourishes in wetland environments is not only important for providing food and shelter, but also for protecting the soil from erosion. The vegetation that thrives in wetlands acts as a buffer zone, protecting the coastline from the effects of nearby storms, tides, and currents and preventing erosion. Furthermore, the presence of various plants promotes the development of bacteria, fungi, and algae, all of which aid in the recycling of nutrients and the decomposition of organic matter. The ecological function and productivity of wetlands both benefit from these activities. They benefit not only the surrounding ecosystem, but also the people who live nearby. They prevent damage from hurricanes and typhoons by acting as natural barriers between the storm and neighbouring settlements. Additionally, they help mitigate climate change by sequestering vast amounts of carbon dioxide in their soil and water. Furthermore, these are important to the well-being of fisheries, provide opportunities for recreation, and enrich the cultural and aesthetic value of a region. Protecting biodiversity and maintaining ecological balance depend critically on the existence of wetland ecosystems. These rare treasures are crucial to the well-being of our planet because of their exceptional characteristics, vital roles in the environment, and abundant populations of life. The survival of these amazing ecosystems for future generations depends on our collective recognition of their value and implementation of effective conservation and restoration practices (Anderson and Davis 2013).

1.1 Importance of Wetland Ecosystems

Incalculable benefits to the natural world, the economy, and human culture make wetland ecosystems rare and valuable resources. They play a crucial role in maintaining a balanced and sustainable ecosystem by filtering water and providing habitat for a wide variety of animal and plant species. The incredible biodiversity of plants and animals in wetland ecosystems makes them some of the most ecologically diverse areas on Earth. Exceptional conditions for the survival of a wide variety of plant and animal species are provided by the interplay between water, soil, and vegetation. These are essential for the survival of many migratory bird, fish, amphibian, reptile, and mammal species because they serve as either primary or secondary breeding, nursery, and feeding grounds. They are home to a large number of unique species and serve as a lifeline for endangered fauna. They rely on their rich diversity of plant communities, which supports a complex web of life. Vegetation in wetlands is used by a wide range of organisms as a source of food, cover, and nesting sites. All types of aquatic plants, even some that can only survive below the surface, are included here. Plants in wetland areas have extensive root systems that improve water quality by stabilising soil, reducing erosion, and capturing sediments and pollutants. Their importance is summarised in Fig. 1.1.

Wetlands play a crucial role in water cycle regulation and hydrological balance maintenance. By absorbing and storing excess precipitation, they operate as natural sponges and lessen the likelihood of floods further downstream during times of heavy precipitation. They collect water during wet times and slowly release it during dry times, providing a consistent supply of water to local ecosystems, human towns, and groundwater reservoirs. They also act as natural filters, purifying water as it passes through them. These are particularly effective at capturing and

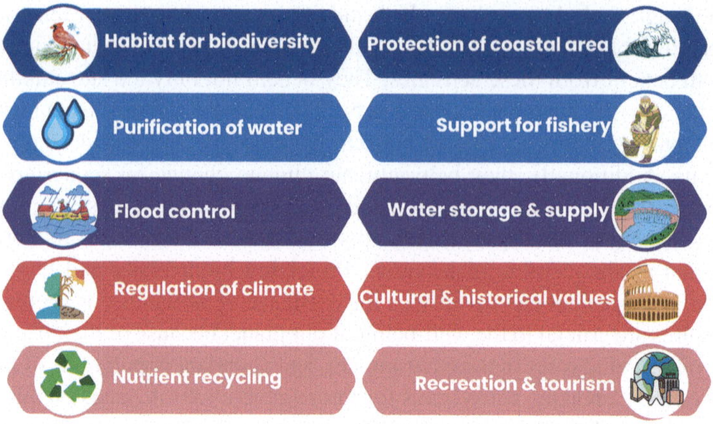

Fig. 1.1 Importance of wetlands

decomposing pollutants, silt, and excess nutrients due to the quantity of dense vegetation and the diversity of microbial communities found there. Water supplies, aquatic life, and ecosystems further downstream all benefit from this filtration process since the water quality is enhanced.

In the fight against global warming, wetland ecosystems are valuable allies to have. They are significant carbon sinks because they remove a lot of CO_2 from the atmosphere and store it. Wetland vegetation in general, and peatlands in particular, are excellent places to store carbon because of their ability to preserve organic materials for thousands of years. When wetlands are drained or degraded, the carbon stored there is released as greenhouse gases, adding to global warming. The amount of carbon emissions produced can be reduced by climate change mitigation efforts including protecting and restoring wetland regions. Adaptation to a changing climate also relies heavily on wetland areas. They act as a natural buffer against the effects of extreme weather events like storms and floods by absorbing and storing excess water. They help moderate temperatures, mitigate the impacts of heat islands, and sustain microclimates that are favourable to many different kinds of life.

Wetlands store water naturally, making them an invaluable resource for human needs including agriculture, industry, and domestic consumption. They are also suitable for sporting and other leisure activities. They are in charge of replenishing groundwater resources, which ensures that residents in the area will always have access to potable water. They are important to commercial fishing because they provide breeding and feeding grounds for a wide diversity of fish species. Millions of people around the world rely heavily on the income generated by wetland resources like fishing, aquaculture, and agricultural production. They also have great value from a recreational and cultural standpoint. In addition to serving as popular tourist attractions for activities like birdwatching, canoeing, and trekking, they also hold religious and cultural value for a large number of indigenous people.

Tourists bring in money and bring attention to environmental issues by visiting wetlands, which is good for the local economy.

The importance of the ecosystems that thrive in wetlands cannot be overstated. Their contributions to biodiversity conservation, water cycle regulation, climate change mitigation, and service supply are invaluable. Recognising the immense value of wetland ecosystems is crucial for their protection and long-term administration. The unique ecological services and benefits provided by wetland habitats must be protected through the implementation of effective conservation practices. Protected areas and other methods of integrated water and land management are examples of such practices. Taking action to protect wetlands and make appropriate use of their resources may help ensure a more secure and resilient future for both nature and humanity (Batzer and Baldwin 2012).

1.2 Role of Wetlands in the Era of Climate Crisis

As the globe grapples with the challenges posed by the climate disaster, wetland ecosystems are emerging as crucial partners in the fight against climate change and the fight to adapt to its repercussions. These unique ecosystems serve a crucial role in climate resilience, carbon sequestration, and water management, among many other important functions. The climate crisis provides an opportunity to examine the many functions that wetlands perform in the modern world.

When it comes to combating global warming, wetland ecosystems are among our best natural friends. They can absorb vast volumes of carbon dioxide from the air and store it indefinitely, making them extremely effective carbon sinks. Plants in wetland habitats, such as marsh grasses and mangroves, collect and store carbon through photosynthesis, while the waterlogged conditions of wetland environments prevent the release of carbon as greenhouse gases. In particular, peatlands play a crucial role in the carbon sequestration process. Over the course of thousands of years, the organic matter in these swampy ecosystems accumulates, resulting in the creation of massive layers of peat. Peatlands contribute to global warming because when they are drained or destroyed, the carbon stored in the peat is liberated as carbon dioxide. To prevent the release of sequestered carbon and to reduce emissions of greenhouse gases, it is crucial to maintain and restore peatlands. In addition, through a process called denitrification, wetland areas aid in reducing emissions of greenhouse gases. Nitrogen, a potent greenhouse gas, is converted into harmless nitrogen gas by microbes in the soils and sediments of wetland environments. This organic cleansing method helps mitigate the damaging impacts of agricultural runoff and reduces Green House Gas (GHG) emissions.

Wetland habitats are crucial for preparing for and adapting to the consequences of climate change. They protect against storms, floods, and droughts by acting as a natural barrier. In the same way, as sponges absorb and store surplus water during heavy rains, so do wetland ecosystems. As a result, fewer downstream floods will be caused. Water is vital to the survival of ecosystems and human settlements, and these structures gently release stored water during dry periods. Because they protect coastlines from erosion, storm surges, and increasing sea levels, coastal wetlands such as mangroves and salt marshes are crucial to climate adaptation. As natural buffers, these ecosystems lessen the destructive potential of coastal storms and mitigate the impact of wave energy. They help keep silt in check and protect habitats for a wide range of animals, all of which contribute to coastal stabilisation. Wetlands are crucial to water resource management and climate adaptation due to their ability to regulate water supply and improve water quality. They act as natural filters, cleaning the water of pollutants and silt so that it is more suitable for consumption before it reaches downstream locations. Recharging groundwater sources through wetland areas is crucial for having a reliable water supply even in dry periods.

Wetland ecosystems are crucial for enhancing the overall ecosystem's resilience to climate change. Its vast floral diversity makes it a suitable habitat for a wide variety of plant and animal species. Ecosystems that have more genetic and species diversity are better able to withstand and recover from disturbances, and this biodiversity helps to strengthen that resilience. The ecosystems that surround wetland habitats also benefit from their presence. They serve as nurseries, feeding grounds, and breeding grounds for numerous species of animals, including economically important fish and migratory birds. The state of landscape health and efficiency is influenced by the degree of interconnectedness between wetland ecosystems and other ecosystems. They also promote the maintenance of healthy soil and water, which are fundamental to the stability of agricultural systems. They help agriculture in several ways, including through the regulation and filtration of water, which results in a higher level of natural moisture in soil and less need for extra water for irrigation. They also offer pollination services, which are critical to the maintenance and growth of agriculturally productive land (Batzer and Sharitz 2014).

Due to the ongoing climate challenge, the significance of wetland ecosystems in the modern day simply cannot be overstated. These exceptional ecosystems play a crucial role in combating climate change, adapting to its effects, and building resilience. Carbon sequestration, water control, and ecosystem health are all areas in which they contribute greatly, making them invaluable in the face of rising environmental issues. Recognising the importance of wetlands and taking measures to protect them is crucial if we are to have a future that is both sustainable and resilient. Wetland management must be incorporated into policies and programmes that address climate change, and conservation and restoration practices must be implemented to increase knowledge about and protect wetland ecosystems. The role of different kinds of wetlands in changing climatic scenarios is given in Table 1.1.

Table 1.1 Role of wetlands in the changing climatic scenario

S. no.	Name of wetland	Country of location	Type of wetland	Role in changing climatic scenario
1	Everglades	USA	Marshland	Carbon storage, water regulation, climate resilience
2	Okavango Delta	Botswana	Inland Delta	Carbon sequestration, flood control, biodiversity preservation
3	Pantanal	Brazil, Bolivia, Paraguay	Seasonal Floodplain	Carbon storage, water purification, climate regulation
4	Sundarbans	Bangladesh, India	Mangrove Forest	Carbon sequestration, coastal protection, climate resilience
5	Camargue	France	Wetland	Carbon storage, biodiversity habitat, flood control
6	Danube Delta	Romania, Ukraine	River Delta	Carbon sequestration, water purification, climate resilience
7	Kakadu National Park	Australia	Wetland	Carbon storage, wildlife habitat, climate adaptation
8	Hula Valley	Israel	Marshland	Carbon sequestration, bird migration, climate regulation
9	Tonle Sap Lake	Cambodia	Flooded Forest	Carbon storage, fisheries, water regulation
10	Lake Chad	Chad, Niger, Nigeria, Cameroon	Endorheic Lake	Climate resilience, water purification, fishing
11	Okefenokee Swamp	USA	Swamp	Carbon sequestration, biodiversity habitat, climate regulation
12	Liaohe Delta	China	River Delta	Carbon storage, coastal protection, climate adaptation
13	Narew National Park	Poland	Wetland	Carbon sequestration, wildlife habitat, water regulation
14	Tisza River Basin	Hungary, Romania, Ukraine	River Basin	Carbon storage, flood control, climate resilience
15	Morava River Floodplain	Serbia, Slovakia, Austria	Floodplain	Carbon sequestration, biodiversity preservation, climate regulation
16	Esteros del Ibera	Argentina	Wetland	Carbon storage, wildlife habitat, water regulation
17	Poyang Lake	China	Seasonal Lake	Carbon sequestration, water purification, flood control
18	Rann of Kutch	India, Pakistan	Seasonal Salt Marsh	Carbon storage, biodiversity habitat, climate adaptation
19	Itaipu Dam Reservoir	Brazil, Paraguay	Reservoir	Climate regulation, water storage, hydroelectric power
20	Humber Estuary	UK	Estuary	Carbon sequestration, flood control, coastal protection

1.3 Organisations Involved in Conservation of Wetlands

Several groups are working on both the national and international levels to ensure the safety of wetland ecosystems. Many more groups are doing important work to protect wetland ecosystems than simply the ones listed here. The efforts of these groups help ensure that these ecosystems are managed and protected in the long term. This assures that wetland habitats, and the benefits they provide to both nature and people, will persist into the future. These groups play a crucial role in promoting sustainable management practices and protecting these ecosystems. The following are some prominent groups that are helping to protect wetlands around the world:

Ramsar Convention: The Ramsar Convention is an international agreement whose stated goal is "the conservation and wise use of the world's wetlands for the benefit of present and future generations". It provides an organisational framework for locating, protecting, and managing globally significant wetlands. There are several countries included in the agreement that have pledged to cooperate in the effort to preserve wetland ecosystems and promote sustainable use of their resources.

Wetlands International: Wetlands International is a global organisation dedicated to preserving and restoring wetland ecosystems. Its goals include studying and protecting wetland ecosystems around the globe, as well as spreading the word about their significance. To safeguard wetlands and put their plans into reality, the organisation collaborates with governments, non-governmental organisations (NGOs), and local people.

The Nature Conservancy: The Nature Conservancy is a pioneering global organisation dedicated to protecting ecosystems like wetland habitats. Wetland areas are prioritised for purchase and management, sustainable land and water management practices are advocated for, and residents are invited to take part in conservation efforts.

World Wildlife Fund (WWF): To protect ecosystems and species diversity on a global scale, WWF is a prominent international organisation. Protecting vital wetland habitats, fostering sustainable water management, and reducing the stresses on wetland species are all goals of the organisation's wetland conservation programmes.

Ducks Unlimited: Ducks Unlimited is a non-profit environmental organisation whose main goal is the preservation of ducks and their natural habitats. It aims to restore and manage wetland areas for the benefit of waterfowl and other animals that rely on these areas for survival. To preserve wetland ecosystems and promote sustainable farming methods, Ducks Unlimited collaborates with private landowners, government agencies, and other organisations.

Environmental NGOs: NGOs dedicated to environmental issues also play a significant role in protecting wetland ecosystems. Environmental advocacy groups such as the Sierra Club, Greenpeace, and Friends of the Earth are trying to protect wetland habitats through direct action, scientific investigation, and engagement with local populations. They spread awareness about the value of wetlands and push for policy changes to ensure their preservation.

Government Agencies: Wetland areas are managed and protected by designated government agencies in several countries. These groups are committed to wetland conservation, restoration, and policy development. Some examples are the US Fish and Wildlife Service, UK's Environment Agency, and Canada's Parks Canada. In India, the Ministry of Environment, Forest and Climate Change (MoEFCC) is located.

1.4 Ramsar Convention

The Ramsar Convention, also known as the Convention on Wetlands, is an international agreement signed by countries all over the world to encourage the conservation and sustainable use of the world's wetlands. The Ramsar Convention on Wetlands of International Importance, adopted in 1971 in the Iranian city of Ramsar, has developed into an international standard for the protection and management of wetlands around the world. The growing recognition of the value of wetland ecosystems and the urgent need to protect them led to the creation of the Ramsar Convention. The worldwide decline and destruction of wetland habitats prompted widespread worry due to urbanisation, agricultural drainage, and other human activities. The Convention set out to address the critical function of wetlands in supporting biodiversity, regulating water cycles, and providing a wide range of ecological services.

The Ramsar Convention aims to achieve the wise use and conservation of wetlands worldwide. Its objectives include:

Halting the loss and degradation of wetlands: The Convention acknowledges the importance of wetland ecosystems for protecting biodiversity and keeping people well and happy, and it stresses the need to safeguard these areas from further degradation and loss.

Promoting the conservation and sustainable use of wetlands: The Convention promotes the sustainable management and use of wetland areas, with due regard for their ecological, economic, cultural, and scientific value.

Establishing a global network of Ramsar sites: Ramsar sites are wetlands that have been officially recognised as being of international significance. The purpose of the Convention is to raise public understanding of the value of protecting wetlands and to inspire the creation and maintenance of examples such as these that showcase effective wetland conservation in action.

The Ramsar Convention relies heavily on the Ramsar sites it designates. They are outstanding examples of wetland habitats that are either environmentally or biologically relevant. These areas serve as "living labs" to test and refine new approaches to wetland conservation and sustainable use. When countries designate areas as Ramsar sites, they gain international recognition for their commitment to protecting and managing the area's wetland ecosystems sustainably. The conservation of biodiversity relies heavily on Ramsar sites. A wide variety of plant and animal species rely on them as critical habitats, including some that are in danger of extinction and others that must travel great distances to get food. These areas are essential stopping points for migrating birds because they give a haven to feed and rest while crossing the country. Additionally, Ramsar sites help maintain the availability of ecosystem services that are vital to human survival. These systems assist in regulating water flows, improve water quality, recharge aquifers, and lessen the impact of droughts and floods.

There have been many significant gains in wetland conservation thanks to the Ramsar Convention's existence. Foregrounding the worldwide awareness and conservation of rich wetland habitats, it has facilitated the designation of hundreds of locations across the globe as Ramsar sites. Leading practices for the management of wetlands and their restoration have been developed thanks in part to the Convention's promotion of international collaboration and the sharing of knowledge. Although this is encouraging, much work remains before the Convention's objectives may be realised. Urbanisation, the expansion of agricultural land, pollution, invasive species, and the consequences of climate change are just some of the threats that wetland ecosystems face. By enhancing national wetland rules, strengthening capacity-building efforts, and encouraging sustainable land-use practices that prioritise the protection of wetland regions, the Convention must find answers to these concerns (Dennison and Berry 1993).

1.5 Global Distributions of Wetlands/Ramsar Sites

Wetland ecosystems are found everywhere on Earth, with each one having its own distinct geography and weather patterns. They cover a considerable portion of Earth's surface and can be found on every continent. Although the following provides a high-level overview of wetland distribution around the world, the specific distribution of wetland types varies widely from region to region. There are currently 2493 sites, which account for a total area of 256,759,538 ha (assessed on 1 June 2023). The following continents feature them, however only to varied degrees.

Africa (424).
Asia (420).
Europe (1126).
Latin America and the Caribbean (216).
North America (222).
Oceania (85).

1.5.1 Africa (424)

Algeria (50), Angola (1), Benin (4), Botswana (1), Burkina Faso (25), Burundi (4), Cabo Verde (4), Cameroon (7), Central African Republic (2), Chad (6), Comoros (3), Congo (14), Côte d'Ivoire (6), Democratic Republic of the Congo (4), Djibouti (1), Egypt (4), Equatorial Guinea (3), Eswatini (3), Gabon (9), Gambia (3), Ghana (6), Guinea (16), Guinea-Bissau (4), Kenya (6), Lesotho (1), Liberia (5), Libya (2),

Madagascar (21), Malawi (2), Mali (4),Mauritania (4), Mauritius (3), Morocco (38), Mozambique (2), Namibia (5), Niger (14), Nigeria (11), Rwanda (1), Sao Tome and Principe (1), Senegal (9), Seychelles (3), Sierra Leone (1), South Africa (29), South Sudan (1), Sudan (4), Togo (4), Tunisia (42), Uganda (12), United Republic of Tanzania (4), Zambia (8), Zimbabwe (7).

1.5.2 Asia (420)

Bahrain (2), Bangladesh (2), Bhutan (3), Cambodia (5), China (82), Democratic People's Republic of Korea (2), India (75), Indonesia (7), Iran (Islamic Republic of) (25), Iraq (4), Japan (53), Jordan (2), Kazakhstan (10), Kuwait (1), Kyrgyzstan (3), Lao People's Democratic Republic (2), Lebanon (4), Malaysia (7), Mongolia (11), Myanmar (6), Nepal (10), Oman (2), Pakistan (19), Philippines (8), Republic of Korea (24), Sri Lanka (6), Syrian Arab Republic (1), Tajikistan (5), Thailand (15), Turkmenistan (1), United Arab Emirates (10), Uzbekistan (3), Viet Nam (9), Yemen (1).

1.5.3 Europe (1126)

Albania (4), Andorra (3), Armenia (3), Austria (23), Azerbaijan (2), Belarus (26), Belgium (9), Bosnia and Herzegovina (3), Bulgaria (11), Croatia (5), Cyprus (1), Czechia (14), Denmark (43), Estonia (17), Finland (49), France (53), Georgia (4), Germany (35), Greece (10), Hungary (29), Iceland (6), Ireland (45), Israel (2), Italy (57), Latvia (6), Liechtenstein (1), Lithuania (7), Luxembourg (2), Malta (2), Monaco (1), Montenegro (3), Netherlands (54), North Macedonia (3), Norway (63), Poland (19), Portugal (31), Republic of Moldova (3), Romania (20), Russian Federation (35), Serbia (11), Slovakia (14), Slovenia (3), Spain (76), Sweden (68), Switzerland (11), Türkiye (14), Ukraine (50), United Kingdom of Great Britain and Northern Ireland (175).

1.5.4 Latin America and the Caribbean (216)

Antigua and Barbuda (1), Argentina (23), Bahamas (1), Barbados (1), Belize (2), Bolivia (Plurinational State of) (11), Brazil (27), Chile (16), Colombia (11), Costa Rica (12), Cuba (6), Dominican Republic (6), Ecuador (19), El Salvador (8), Grenada (1), Guatemala (7), Honduras (12), Jamaica (4), Nicaragua (9), Panama (5), Paraguay (6), Peru (14), Saint Lucia (2), Suriname (1), Trinidad and Tobago (3), Uruguay (3), Venezuela (Bolivarian Republic of) (5).

1.5.5 North America (222)

Canada (37), Mexico (144), United States of America (41).

1.5.6 Oceania (85)

Australia (67), Fiji (2), Kiribati (1), Marshall Islands (2), New Zealand (7), Palau (1), Papua New Guinea (2), Samoa (2), Vanuatu (1).

1.6 Classification of Wetlands

It's important to note that wetlands can often exhibit characteristics of multiple classifications, and the classification systems can vary depending on regional or scientific contexts. These classifications provide a framework for understanding and studying the diverse range of wetland ecosystems found worldwide. Wetlands can be classified based on various criteria, including their location, hydrology, vegetation, and ecological characteristics as given in Fig. 1.2. Here are some common classification systems used for wetlands:

1.6.1 Location-Based Classification

Coastal Wetlands: Wetland habitats such as mangrove forests, salt marshes, and coastal lagoons are characteristic of coastal areas because they are moulded by the presence of tidal or marine waters.

Inland Wetlands: Swamps, marshes, bogs, fens, and floodplains are all examples of inland wetlands, as are saline lakes like salt pans.

1.6.2 Hydrological Classification

Riverine Wetlands: Riparian zones, floodplains, and oxbow lakes are all types of wetlands found near rivers.

Lacustrine Wetlands: Lake and reservoir-linked wetland habitats, including deltas, backwaters, and shallow lakes.

Fig. 1.2 Classification of wetlands

Palustrine Wetlands: Swamps, marshes, bogs, and fens are all examples of wetland types that can be found in depressions or basins. Shallow standing water is a common feature of these wetland ecosystems.

1.6.3 Vegetation-Based Classification

Forested Wetlands: Swamps and forests made up primarily of woody plants are called wet forest ecosystems.

Herbaceous Wetlands: Herbaceous plants such as grasses, sedges, and rushes predominate in these wetlands. Marshes, meadows, and prairie potholes are all examples of such ecosystems.

1.6.4 Geological Classification

Peatlands: Peatlands are wetland ecosystems where partially decomposed organic matter (peat) has accumulated. Bogs, fens, and mires are all examples of peatlands.

Tidal Wetlands: Salt marshes and tidal flats are examples of wetlands created by the tides.

1.6.5 Ecological Classification

Lotic Wetlands: Streams and wetlands that are connected to moving water, such as those found along rivers.

Lentic Wetlands: Lakes, ponds, and low-lying wetland areas are all examples of wetland types linked with standing or slowly moving water.

1.6.6 Human-Made Wetlands

Constructed Wetlands: Man-made wetlands provide a variety of functions, including cleaning up wastewater, controlling flooding, and restoring wildlife habitat.

1.6.6.1 According to Ramsar Convention

This category refers specifically to wetlands designated as Ramsar Sites under the Ramsar Convention. These sites are recognised for their international importance in terms of biodiversity, ecological significance, or cultural value.

Marine/Coastal Wetlands

A—Permanent shallow marine waters in most cases less than six metres deep at low tide; include sea bays and straits.

B—Marine subtidal aquatic beds; include kelp beds, sea-grass beds, and tropical marine meadows.
C—Coral reefs.
D—Rocky marine shores; include rocky offshore islands and sea cliffs.
E—Sand, shingle, or pebble shores; include sand bars, spits, and sandy islets; and dune systems and humid dune slacks.
F—Estuarine waters; permanent water of estuaries and estuarine systems of deltas.
G—Intertidal mud, sand, or salt flats.
H—Intertidal marshes; include salt marshes, salt meadows, saltings, and raised salt marshes; includes tidal brackish and freshwater marshes.
I—Intertidal forested wetlands; include mangrove swamps, nipah swamps, and tidal freshwater swamp forests.
J—Coastal brackish/saline lagoons; brackish to saline lagoons with at least one relatively narrow connection to the sea.
K—Coastal freshwater lagoons; include freshwater delta lagoons.
Zk(a)—Karst and other subterranean hydrological systems, marine/coastal.

Inland Wetlands
L—Permanent inland deltas.
M—Permanent rivers/streams/creeks; includes waterfalls.
N—Seasonal/intermittent/irregular rivers/streams/creeks.
O—Permanent freshwater lakes (over 8 ha); includes large oxbow lakes.
P—Seasonal/intermittent freshwater lakes (over 8 ha); include floodplain lakes.
Q—Permanent saline/brackish/alkaline lakes.
R—Seasonal/intermittent saline/brackish/alkaline lakes and flats.
Sp—Permanent saline/brackish/alkaline marshes/pools.
Ss—Seasonal/intermittent saline/brackish/alkaline marshes/pools.
Tp—Permanent freshwater marshes/pools; ponds (below 8 ha), marshes, and swamps on inorganic soils; with emergent vegetation water-logged for at least most of the growing season.
Ts—Seasonal/intermittent freshwater marshes/pools on inorganic soils; includes sloughs, potholes, seasonally flooded meadows, and sedge marshes.
U—Non-forested peatlands; include shrub or open bogs, swamps, and fens.
Va—Alpine wetlands; include alpine meadows, and temporary waters from snowmelt.
Vt—Tundra wetlands; include tundra pools, and temporary waters from snowmelt.
W—Shrub-dominated wetlands; shrub swamps, shrub-dominated freshwater marshes, shrub carralder thicket on inorganic soils.
Xf—Freshwater, tree-dominated wetlands; include freshwater swamp forests, seasonally flooded forests, and wooded swamps on inorganic soils.
Xp—Forested peatlands; peat swamp forests.
Y—Freshwater springs; oases.
Zg—Geothermal wetlands.
Zk(b)—Karst and other subterranean hydrological systems, inland.

Human-Made Wetlands

1—Aquaculture (e.g., fish/shrimp) ponds.
2—Ponds; include farm ponds, stock ponds, and small tanks (generally below 8 ha).
3—Irrigated land; includes irrigation channels and rice fields.
4—Seasonally flooded agricultural land (including intensively managed or grazed wet meadow or pasture).
5—Salt exploitation sites; salt pans, salines, etc.
6—Water storage areas; reservoirs/barrages/dams/impoundments (generally over 8 ha).
7—Excavations; gravel/brick/clay pits; borrow pits, mining pools.
8—Wastewater treatment areas; sewage farms, settling ponds, oxidation basins, etc.
9—Canals and drainage channels, ditches.
Zk(c)—Karst and other subterranean hydrological systems, human-made.

1.7 Ecosystem Services

The term "ecosystem services" refers to the advantages that people get as a result of healthy ecosystems. All or part of these advantages originate in the world outside. These services are important to our survival because they ensure our access to a wide variety of goods and assistance that are vital to our ability to earn a living and stay healthy. Ecosystem services are the vital benefits we obtain as a result of healthy natural ecosystems. Ecosystem services are the basis for all other human activities, and include cultural services like recreation and aesthetic value, regulatory services like flood management and climate regulation, provisioning services like food and raw materials, and sustaining services like carbon sequestration. The importance of ecosystem services must be recognised if we are to achieve sustainable development and safeguard our natural environment. If these services are recognised for their worth and safeguarded, future generations will be able to live in peace and prosperity.

1.7.1 Importance of Ecosystem Services

Ecosystem services are essential to the well-being and survival of both human societies and the natural environment. They provide numerous advantages necessary for maintaining our standard of living, society's economic foundations, and the planet's ecological balance some of which are given in Table 1.2. The following are examples of the importance of ecosystem services that are particularly noteworthy:

Human Well-Being: Examples of ecosystem services that directly benefit human well-being include the provision of food, clean water, clean air, and materials for housing, clothing, and medicine. They also offer opportunities for leisure and pleasure, beauty and enlightenment, and cultural and spiritual growth. Ecosystems provide essential support for human well-being, including physical, mental, and emotional health, and quality of life.

Table 1.2 Examples of wetlands and their ecosystem services

S. no.	Name of wetland	Country of location	Type of wetland	Ecosystem services provided
1	Everglades	USA	Freshwater Wetland	Flood control, water purification, wildlife habitat
2	Sundarbans	Bangladesh, India	Mangrove Wetland	Storm protection, fish breeding, carbon sequestration
3	Okavango Delta	Botswana	Inland Delta Wetland	Biodiversity, water filtration, tourism
4	Pantanal	Brazil, Bolivia	Flooded Grassland	Biodiversity hotspot, water purification, ecotourism
5	Camargue	France	Salt Marsh Wetland	Bird habitat, flood protection, salt production
6	Kakadu National Park	Australia	Freshwater Wetland	Cultural significance, biodiversity, tourism
7	Ciénaga Grande de Santa Marta	Colombia	Coastal Wetland	Fisheries, water purification, bird habitat
8	Danube Delta	Romania, Ukraine	River Delta Wetland	Bird migration, fisheries, carbon sequestration
9	Okefenokee Swamp	USA	Freshwater Wetland	Wildlife habitat, recreation, water storage
10	Wadden Sea	Germany, Netherlands	Coastal Wetland	Coastal protection, bird habitat, tourism
11	Hula Lake	Israel	Freshwater Wetland	Bird migration, water purification, recreation
12	Tonle Sap Lake	Cambodia	Seasonal Floodplain	Fisheries, rice production, bird habitat
13	Lake Baikal	Russia	Freshwater Wetland	Biodiversity, water purification, tourism
14	Térraba-Sierpe National Wetlands	Costa Rica	Mangrove Wetland	Carbon sequestration, fish nursery, water filtration
15	Chilika Lake	India	Brackish Water Lagoon	Fishery, bird habitat, aquaculture
16	Tana River Delta	Kenya	River Delta Wetland	Biodiversity, water purification, agriculture
17	Lagoa do Peixe	Brazil	Coastal Lagoon	Bird migration, tourism, water purification
18	Lake Titicaca	Peru, Bolivia	Freshwater Lake	Fisheries, tourism, water source
19	Rann of Kutch	India, Pakistan	Desert Salt Marsh	Wildlife habitat, salt production, tourism
20	Chokpak Wetland	Kazakhstan	Saline Wetland	Bird migration, habitat for wildlife, research

Supporting Life and Biodiversity: The existence and diversity of life on Earth would not be possible without the services offered by ecosystems. They provide a habitat in which plants, animals, and microorganisms can thrive, reproduce, and interact. In return, biodiversity helps ecosystems to be more robust, stable, and adaptable to environmental change.

Climate Regulation: Climate regulation relies heavily on the services provided by ecosystems. Ecosystems high in vegetation and capable of absorbing and storing carbon dioxide are known as carbon sinks. Carbon dioxide (CO_2) is a greenhouse gas that exacerbates global warming. Areas that are rich in plants, such as forests and marshes. They also regulate the weather, making the environment more habitable by stabilising temperatures, humidity levels, and rainfall patterns.

Water Regulation: Ecosystem services control the water cycle, which keeps freshwater supplies available and helps to maintain their quality. The sediments, minerals, and contaminants in the water that pass through wetlands, forests, and other habitats are naturally filtered out. The risk of flooding, drought, and water scarcity is mitigated because of their ability to control water flow.

Soil Fertility and Nutrient Cycling: Ecosystem services contribute to soil fertility and the cycling of nutrients. Plants, microorganisms, and decomposers help break down organic matter, releasing nutrients that are essential for plant growth. Healthy soils, in turn, support agricultural productivity, food security, and sustainable land use practices.

Pollination and Pest Control: Ecosystem services like the work of pollinators like bees, butterflies, and birds, are essential to the reproduction of flowering plants and the growth of edible crops like fruits, vegetables, and seeds. There is less need for chemical pesticides and more support for the use of natural pest control approaches since predatory creatures serve to regulate pest populations within ecosystems.

Erosion Control: Help to the maintenance of soil quality including vegetation and plant roots, which act as buffers against the destructive forces of wind and water. They prevent sediment from entering waterways and stabilise the soil, both of which are important for farming.

Economic Value and Livelihoods: Ecosystem services have a significant monetary value and are crucial to the upkeep of many different types of economies. They contribute significantly to the food and beverage, forestry, travel and leisure, and healthcare sectors. The delivery of ecosystem services has been linked to the development of economic opportunities, particularly in rural and coastal areas.

Climate Change Adaptation and Resilience: Ecosystems and human cultures may be better able to adjust to climate change if ecosystem services are provided. When in good condition, ecosystems such as mangroves, coral reefs, and forests can protect coastlines from storms, erosion, and increasing sea levels. In addition, they back up initiatives to adapt to climate change, like ecosystem-based water management, risk reduction, and sustainable land use.

Sustainable Development: SDGs cannot be reached without the assistance of ecosystem services. They provide a foundation for sustainable development's three pillars—social, economic, and environmental—and thereby promote equity, resilience, and long-term well-being for today's and tomorrow's generations.

1.7.2 Types of Ecosystem Services

Ecosystem services are the benefits provided by ecosystems to people. Without these services, our health, economy, and quality of life would suffer significantly. Ecosystem services can be roughly classified into the following four broad classes as shown in Fig. 1.3 and explained in detail in the sections below.

Provisioning Services Ecosystems provide a wide variety of services, including the provision of direct products and resources including food, water, timber, fibre, fuel, and medicinal plants. Such benefits are made possible by the work of ecosystems. Wetlands provide many benefits, including food (in the form of fish and shellfish), water (for irrigation or drinking), and building materials (in the form of reeds and sedges) which are also mentioned in Table 1.3. The most frequently encountered provisioning services are as follows.

Food: Crops, fruits, vegetables, nuts, and seafood are just a few of the many food sources that are made possible by ecosystems. Agricultural ecosystems, such as farmland and orchards, are essential to the creation of food and other agricultural goods. The wild vegetables, mushrooms, and fish that can be found in places like forests and swamps constitute an important part of the global food supply.

Water: Because they control the quantity, quality, and flow of water, ecosystems play a crucial role in the water supply process. They do this by relying on natural processes including groundwater recharge, water purification, and the upkeep of water retention in bodies of water like lakes, rivers, and wetlands. This contributes to the security of an adequate water supply. Ecosystems are also important because they help ensure sufficient supplies of freshwater for human and agricultural needs.

Timber and Fiber: Industries as diverse as building, furniture making, and paper production rely on forest ecosystems for essential commodities like lumber, wood, and fibre. By employing sustainable forest management practices, we can secure the continued availability of these materials while lessening our ecological footprint.

Fig. 1.3 Types of ecosystem services

Table 1.3 Examples of provisioning services provided by wetlands

S. no.	Name of wetland	Country of location	Type of wetland	Provisioning ecosystem services provided
1	Everglades	USA	Freshwater Wetland	Fishery, water supply, reed for thatch and crafts
2	Sundarbans	Bangladesh, India	Mangrove Wetland	Fishery, honey production, timber
3	Okavango Delta	Botswana	Inland Delta Wetland	Fishery, water supply, thatch for roofing
4	Pantanal	Brazil, Bolivia	Flooded Grassland	Fishery, cattle grazing, water supply
5	Camargue	France	Salt Marsh Wetland	Salt production, fishery, cattle grazing
6	Kakadu National Park	Australia	Freshwater Wetland	Fishery, hunting, gathering of edible plants
7	Ciénaga Grande de Santa Marta	Colombia	Coastal Wetland	Fishery, crab harvesting, agriculture
8	Danube Delta	Romania, Ukraine	River Delta Wetland	Fishery, reed harvesting, water supply
9	Okefenokee Swamp	USA	Freshwater Wetland	Alligator harvesting, water supply, hunting
10	Wadden Sea	Germany, Netherlands	Coastal Wetland	Fishery, shellfish harvesting, cattle grazing
11	Hula Lake	Israel	Freshwater Wetland	Fishery, papyrus for paper, water supply
12	Tonle Sap Lake	Cambodia	Seasonal Floodplain	Fishery, rice cultivation, floating villages
13	Lake Baikal	Russia	Freshwater Wetland	Fishery, water supply, medicinal plants
14	Térraba-Sierpe National Wetlands	Costa Rica	Mangrove Wetland	Fishery, shrimp farming, timber
15	Chilika Lake	India	Brackish Water Lagoon	Fishery, prawn farming, reed for roofing
16	Tana River Delta	Kenya	River Delta Wetland	Fishery, agriculture, water supply
17	Lagoa do Peixe	Brazil	Coastal Lagoon	Fishery, shrimp farming, cattle grazing
18	Lake Titicaca	Peru, Bolivia	Freshwater Lake	Fishery, totora reed for crafts, aquaculture
19	Rann of Kutch	India, Pakistan	Desert Salt Marsh	Salt production, fishery, cattle grazing
20	Chokpak Wetland	Kazakhstan	Saline Wetland	Fishery, cattle grazing, medicinal plants

Medicinal Plants: Many plant species found in ecosystems have qualities that can be used for medical purposes. Many cultures' traditional medicines are derived from plants that grow in forests, swamps, and grasslands. Plants with therapeutic properties are a vital source of materials for the pharmaceutical industry, which in turn improves people's health and well-being.

Fuel and Energy: Ecosystems provide numerous potential sources of fuel and energy, including wood, biomass, and a wide range of biofuels. Biomass fuels are used for both heating and cooking in many communities, especially those in rural areas. Practices like sustainable management of forest resources and the development of renewable energy sources can aid in the provision of electricity that is both clean and reliable.

Genetic Resources: Ecosystems operate as a repository for the genetic diversity of wild plant and animal species. New plant varieties, livestock breeds, and agricultural and fishery advances can't be achieved without this genetic variety. It also contributes to the funding of scientific research, biotechnology, and the development of new medicines.

Raw Materials: Ecosystems provide an abundant source of raw materials including minerals, metals, and other chemical compounds. The availability of these resources is crucial to the success of mining and other resource extraction activities, which in turn benefits the economy and keeps industry running smoothly.

Ecotourism and Recreation: Having more opportunities for tourism and recreation is one benefit of ecosystems' popularity as travel destinations. Natural landscapes, national parks, and protected zones are great places to go hiking, wildlife watching, camping, and engaging in other forms of outdoor recreation. Ecotourism boosts local economies while also assisting in the protection of ecological systems (Denyer and Peters 2022; Finlayson et al. 2018a, b).

Regulating Services These functions include maintaining the current environmental conditions and regulating the processes that take place within an ecosystem. As natural sponges, wetlands play a crucial role in the process of water flow control by absorbing and storing excess water during floods and slowly releasing it during droughts. They also serve as natural water filters and purifiers, removing impurities while improving water quality. Regulating ecosystem services are those that assist in keeping ecological processes in check and stable which are mentioned in Table 1.4. Standard examples of regulatory ecosystem services include the ones listed below.

Climate Regulation: One of the most important ways in which ecosystems help to regulate the climate is by sequestering carbon dioxide through photosynthesis and storing it in biomass and soils. Carbon sinks are ecosystems with a lot of vegetation that help reduce the consequences of global warming. Forests and wetlands are two examples of these types of habitats. They also control regional and local climate by adjusting factors including humidity, temperature, and precipitation.

Water Regulation: The interaction between ecosystems and their parts is a major determinant of the water cycle. In order to accomplish this, they perform the functions including evapotranspiration, groundwater recharge, and runoff management, from which they finally release water. This aids in managing the water supply. Water quality is enhanced by the natural filtration provided by wetlands, forests, and riparian zones, which trap sediments, nutrients, and contaminants also.

Table 1.4 Examples of regulating services provided by wetlands

S. no.	Name of wetland	Country of location	Type of wetland	Regulating ecosystem services provided
1	Everglades	USA	Freshwater Wetland	Flood regulation, water purification, climate regulation
2	Sundarbans	Bangladesh, India	Mangrove Wetland	Storm protection, water quality improvement, erosion control
3	Okavango Delta	Botswana	Inland Delta Wetland	Flood control, water filtration, carbon sequestration
4	Pantanal	Brazil, Bolivia	Flooded Grassland	Flood mitigation, water purification, carbon storage
5	Camargue	France	Salt Marsh Wetland	Coastal protection, water purification, erosion prevention
6	Kakadu National Park	Australia	Freshwater Wetland	Flood control, water quality improvement, carbon storage
7	Ciénaga Grande de Santa Marta	Colombia	Coastal Wetland	Storm surge protection, water quality improvement, erosion control
8	Danube Delta	Romania, Ukraine	River Delta Wetland	Flood regulation, water purification, sediment retention
9	Okefenokee Swamp	USA	Freshwater Wetland	Flood control, water quality improvement, carbon sequestration
10	Wadden Sea	Germany, Netherlands	Coastal Wetland	Coastal flood regulation, water purification, erosion control
11	Hula Lake	Israel	Freshwater Wetland	Flood control, water quality improvement, climate regulation
12	Tonle Sap Lake	Cambodia	Seasonal Floodplain	Flood regulation, water purification, sediment retention
13	Lake Baikal	Russia	Freshwater Wetland	Water temperature regulation, water purification, climate regulation
14	Térraba-Sierpe National Wetlands	Costa Rica	Mangrove Wetland	Storm protection, water quality improvement, erosion control
15	Chilika Lake	India	Brackish Water Lagoon	Storm surge protection, water quality improvement, sediment retention
16	Tana River Delta	Kenya	River Delta Wetland	Flood regulation, water purification, erosion prevention
17	Lagoa do Peixe	Brazil	Coastal Lagoon	Coastal protection, water purification, climate regulation
18	Lake Titicaca	Peru, Bolivia	Freshwater Lake	Flood control, water quality improvement, carbon storage
19	Rann of Kutch	India, Pakistan	Desert Salt Marsh	Storm protection, water quality improvement, erosion control
20	Chokpak Wetland	Kazakhstan	Saline Wetland	Flood mitigation, water purification, sediment retention

Flood Control: Ecosystems including wetlands, floodplains, and mangrove forests act as natural flood barriers. They can hold water for later use, slow the rate of runoff, and soak up excess precipitation, all of which lessen the likelihood of flooding. The ability of these ecosystems to moderate flood levels protects human communities and other built environments.

Erosion Control: Ecosystems including woods, meadows, and plants along the coast help reduce soil erosion. Plant roots stabilise the soil, reducing the likelihood of landslides and erosion caused by water or wind. By preventing soil erosion and keeping the soil in its original structure, ecosystems protect fertile farmland and reduce the quantity of silt that is washed into waterways.

Pollination and Pest Control: Ecosystems have an important role in pollination, which in turn ensures the survival of blooming plant species and the production of fruits, vegetables, and seeds. Pollinating insects, birds, bats, and other creatures play a crucial role in this process. The presence of predatory insects, birds, and other species provides a form of natural pest management, and ecosystems provide this service as well. The use of chemical pesticides can be minimised thanks to the beneficial effects of these organisms.

Disease Regulation: Disease rates and outcomes can be modulated by ecosystems. When natural areas are left untouched, such as forests and wetlands, they help maintain a balance between disease-causing organisms and the hosts they infect. They may prevent the spread of disease by acting as a natural barrier and reducing the number of vector-borne illnesses.

Air Quality Regulation: The removal of pollutants and the creation of oxygen via photosynthesis are only two of the ways in which ecosystems boost air quality. Improved air quality and human health result from plants, especially forests, being able to take in and process harmful substances including carbon monoxide, nitrogen oxides, and particle matter.

Noise and Visual Pollution Reduction: Ecosystem preservation, in particular of urban green spaces and natural landscapes, may reduce noise pollution and increase natural beauty. Vegetation acts as a sound barrier, absorbing and deflecting sounds, while natural landscapes provide aesthetically pleasing surroundings that improve well-being and quality of life (Gearey et al. 2020; Giblett 2019).

Supporting Services These services, which include things like nutrient cycling, soil formation, and plant pollination, are necessary for the production of other ecosystem services. Wetlands not only help with nutrient cycling by absorbing and storing nutrients in the soils and vegetation of the places where they are located, but they also provide a habitat for a broad range of plant and animal species, including pollinators that are crucial to agricultural output. The most essential services provided by ecosystems are those that ensure the survival of other services and the smooth functioning of ecosystems as a whole; these are called supporting ecosystem services. They are crucial to the well-being of both humans and the environment because they ensure the continuity of ecological processes essential to the survival of all forms of life. Some of them are explained in Table 1.5. Common ecological services include the ones listed below.

Table 1.5 Examples of supporting services provided by wetlands

S. no.	Name of wetland	Country of location	Type of wetland	Supporting services ecosystem services provided
1	Everglades	USA	Freshwater Wetland	Habitat for aquatic life, nutrient cycling, primary production
2	Sundarbans	Bangladesh, India	Mangrove Wetland	Nursery for marine species, habitat for wildlife, sediment trapping
3	Okavango Delta	Botswana	Inland Delta Wetland	Breeding grounds for fish and birds, habitat for mammals, nutrient cycling
4	Pantanal	Brazil, Bolivia	Flooded Grassland	Biodiversity hotspot, feeding grounds for migratory birds, nutrient cycling
5	Camargue	France	Salt Marsh Wetland	Bird habitat, shelter for fish and invertebrates, nutrient cycling
6	Kakadu National Park	Australia	Freshwater Wetland	Habitat for waterbirds, nutrient cycling, primary production
7	Ciénaga Grande de Santa Marta	Colombia	Coastal Wetland	Fish nursery, habitat for migratory birds, nutrient cycling
8	Danube Delta	Romania, Ukraine	River Delta Wetland	Bird nesting and feeding grounds, fish habitat, sediment trapping
9	Okefenokee Swamp	USA	Freshwater Wetland	Habitat for reptiles and amphibians, nutrient cycling, water storage
10	Wadden Sea	Germany, Netherlands	Coastal Wetland	Stopover for migratory birds, breeding grounds for fish, nutrient cycling
11	Hula Lake	Israel	Freshwater Wetland	Bird habitat, fish breeding grounds, nutrient cycling
12	Tonle Sap Lake	Cambodia	Seasonal Floodplain	Fish spawning, breeding ground for birds, nutrient cycling
13	Lake Baikal	Russia	Freshwater Wetland	Biodiversity, habitat for endemic species, nutrient cycling
14	Térraba-Sierpe National Wetlands	Costa Rica	Mangrove Wetland	Nursery for marine species, habitat for reptiles, nutrient cycling
15	Chilika Lake	India	Brackish Water Lagoon	Bird habitat, fish breeding, nutrient cycling
16	Tana River Delta	Kenya	River Delta Wetland	Fish nursery, bird habitat, sediment trapping
17	Lagoa do Peixe	Brazil	Coastal Lagoon	Bird migration stopover, fish nursery, nutrient cycling
18	Lake Titicaca	Peru, Bolivia	Freshwater Lake	Fish breeding, habitat for waterbirds, nutrient cycling
19	Rann of Kutch	India, Pakistan	Desert Salt Marsh	Bird habitat, shelter for wildlife, nutrient cycling
20	Chokpak Wetland	Kazakhstan	Saline Wetland	Bird habitat, habitat for migratory species, nutrient cycling

Biodiversity Support: Ecosystems are crucial to the maintenance of biodiversity because they provide a home for a wide range of plant, animal, and microbial species. The ability of an ecosystem to function, be resilient, and adapt to changing environmental conditions is all dependent on its biodiversity. It is extremely important for the stability of the ecosystem, the cycle of nutrients, and the relationships between different species.

Nutrient Cycling: Ecosystems allow for the cycling and recycling of nutrients like carbon, nitrogen, phosphorus, and many other crucial components. Organic debris is broken down by decomposers like bacteria and fungi, which then release nutrients that plants may absorb. Primary producers' access to the nutrients they need to survive is maintained through nutrient cycling, which in turn promotes the growth of plants and other animals.

Soil Formation and Maintenance: Ecosystems provide essential assistance in soil formation and management. Ecosystems contribute to soil formation by, for example, hastening the breakdown of organic matter and weathering of rocks. Plants' roots link soil particles together, increasing the soil's stability, while vegetation itself helps to minimise soil erosion by reducing the force exerted by wind and water.

Primary Production: Photosynthesis, the process by which plants and other photosynthetic organisms convert the energy from sunlight into organic compounds rich in energy, is crucial to the continued health of ecosystems. This output is crucial to the development and growth of other organisms in the environment, and it forms the basis of food chains.

Habitat Provision: An ecosystem provides shelter and food for a wide variety of animal species. Forests, grasslands, wetlands, and coral reefs are just some of the habitats that provide ideal circumstances for a wide range of plant and animal species. These settings facilitate species variety, population mobility, and the stability of ecological communities.

Genetic Resources: Maintaining species' present genetic diversity depends on the availability of genetic resources throughout ecosystems. This genetic diversity is crucial for species to adapt, be resilient, and persist in dynamic environments. Utilising genetic resources helps further scientific inquiry, breeding efforts, and the development of novel food, fibre, and medicine products.

Water Filtration: Wetland and riparian ecosystems in particular act as natural filters, cleaning water of sediments, toxins, and excess nutrients. Improved water quality and the continued vitality of aquatic ecosystems are benefits of these habitats' plant and soil participation in the capture and transformation of contaminants.

Carbon Sequestration: One of the most significant ways ecosystems may help mitigate climate change is by sequestering carbon. Through a process called photosynthesis, plants are able to pull carbon dioxide out of the air and store it in their biomass and the soil. Ecosystems that store large quantities of carbon and hence reduce greenhouse gas concentrations are known as carbon sinks. Forests, grasslands, and other vegetation-rich environments are all good examples of carbon sinks (Kar 2013; Keddy 2010; Kent 2000).

Cultural Services Ecosystem services are the immeasurable advantages to human well-being that may be directly attributed to natural environments. Opportunities for learning and cultural enrichment, as well as leisure pursuits, are only a few examples. Spiritual rites, ancient traditions, and recreational activities like birdwatching, fishing, and boating all take place in wetland settings, giving these areas a high level of cultural value for many cultures around the world. In addition, they serve as a teaching tool for classes dealing with ecological systems and biodiversity. The intangible benefits that people obtain from ecosystems are known as "cultural ecosystem services". Ecosystems enrich our lives in many ways, including culturally, spiritually, and via opportunities for pleasure. These services have important implications for people's well-being and ability to enjoyment of the outdoors. Some of the examples are given in Table 1.6. Some typical cultural ecosystem services are as follows:

Recreation and Tourism: Tourism and recreation could both benefit from the expansion of ecosystems. Thanks to parks, beaches, forests, and other beautiful areas, this outdoors can be a wonderful place to go hiking, camping, birdwatching, and photography. In addition to improving people's physical and mental health, these pastimes allow them to unwind and appreciate nature (Křeček et al. 2006).

Aesthetic and Inspirational Value: Aesthetic value can be attributed to ecosystems since they are the driving force behind the development of visually pleasant landscapes. Waterfalls, mountains, coastlines, and ecosystems that are one-of-a-kind inspire feelings of awe, thankfulness, and amazement in individuals. Poetry, music, literature, and painting are just a few of the many art forms that have been sparked by nature's amazing beauty.

Spiritual and Cultural Significance: Ecosystems hold significant spiritual and cultural meaning for many populations. Sacred sites, cultural landscapes, and heritage sites are often associated with specific ecosystems. These spots are ideal for doing spiritual or religious rites and ceremonies and for cultivating a sense of community and kinship with one's ancestors.

Education and Research: Ecosystems are crucial in fostering learning, scientific investigation, and a broader understanding of the natural world. They serve as dynamic learning environments, rich with opportunities for scientific inquiry and hands-on experience. Ecosystems underpin the growth of scientific knowledge and efforts to maintain biodiversity through the study of biodiversity, ecological research, and environmental monitoring.

Sense of Place and Identity: Both our sense of place and our sense of self are formed by the ecosystems in which we live. The local community's history, traditions, and expertise are intrinsic to them. For example, indigenous tribes have deep historical links to specific biomes. Indigenous peoples' identities, spiritualities, and customs all stem from these interconnected spheres of life.

Sense of Well-Being and Health: Ecosystems contribute positively to human well-being and health by, among other things, providing places to unwind, relieve stress, and get some exercise. Spending time in natural settings has been associated with positive mental health outcomes, including decreased stress and enhanced well-being. Ecosystems also facilitate healthy lifestyles by providing venues for physical activity, sports, and relaxation.

Table 1.6 Examples of cultural services provided by wetlands

S. no.	Name of wetland	Country of location	Type of wetland	Cultural services ecosystem services provided
1	Everglades	USA	Freshwater Wetland	Recreation, tourism, cultural significance, education
2	Sundarbans	Bangladesh, India	Mangrove Wetland	Tourism, spiritual and religious significance, traditional practices
3	Okavango Delta	Botswana	Inland Delta Wetland	Tourism, cultural heritage, indigenous knowledge
4	Pantanal	Brazil, Bolivia	Flooded Grassland	Ecotourism, traditional fishing, local livelihoods
5	Camargue	France	Salt Marsh Wetland	Birdwatching, photography, cultural events, festivals
6	Kakadu National Park	Australia	Freshwater Wetland	Aboriginal heritage, cultural education, storytelling
7	Ciénaga Grande de Santa Marta	Colombia	Coastal Wetland	Ecotourism, cultural practices, folklore, traditional knowledge
8	Danube Delta	Romania, Ukraine	River Delta Wetland	Birdwatching, boat tours, cultural festivals, traditional crafts
9	Okefenokee Swamp	USA	Freshwater Wetland	Wildlife photography, storytelling, canoeing, cultural heritage
10	Wadden Sea	Germany, Netherlands	Coastal Wetland	Birdwatching, hiking, local traditions, coastal culture
11	Hula Lake	Israel	Freshwater Wetland	Birdwatching, environmental education, cultural events, folklore
12	Tonle Sap Lake	Cambodia	Seasonal Floodplain	Cultural tourism, boat trips, traditional fishing practices
13	Lake Baikal	Russia	Freshwater Wetland	Cultural tourism, indigenous communities, folklore
14	Térraba-Sierpe National Wetlands	Costa Rica	Mangrove Wetland	Ecotourism, cultural exchange, traditional practices
15	Chilika Lake	India	Brackish Water Lagoon	Birdwatching, cultural festivals, fishing traditions
16	Tana River Delta	Kenya	River Delta Wetland	Birdwatching, cultural tours, indigenous knowledge
17	Lagoa do Peixe	Brazil	Coastal Lagoon	Birdwatching, cultural events, local traditions
18	Lake Titicaca	Peru, Bolivia	Freshwater Lake	Cultural tourism, indigenous communities, traditional practices
19	Rann of Kutch	India, Pakistan	Desert Salt Marsh	Cultural festivals, traditional crafts, folk music
20	Chokpak Wetland	Kazakhstan	Saline Wetland	Birdwatching, cultural exchange, traditional practices

Sense of Resilience and Adaptation: Resilience and the ability to adjust to new conditions can be fostered within an ecosystem. Learning from natural ecosystems and drawing inspiration from them through observation and involvement can promote adaptability and openness to change. Ecosystems excel at demonstrating sustainable practices and are fantastic examples of systems that are resilient enough to coexist with nature.

Cultural Practices and Traditions: To maintain cultural practices, indigenous knowledge, and conventional uses of natural resources, ecosystems are crucial. Fishing, hunting, and gathering wild foods are all examples of traditional agricultural practices that are deeply rooted in the ecosystems in which they occur. These efforts aid in the safeguarding of cultural artefacts, the maintenance of cultural traditions, and the transmission of information (Marble 1991).

1.8 Threats to Wetland Ecosystem

When an object or event has the potential to cause harm, damage, or danger to other people, we call it a threat. In the context of ecosystems and the environment, a threat is anything that endangers the well-being, stability, or efficiency of these natural systems. These threats can emerge from a variety of sources, including human activity, natural disasters, and anthropogenic changes to the environment. Human activity can negatively affect ecosystems in a wide variety of ways, including loss of habitat, pollution, climate change, invasive species, overexploitation of resources, and unsustainable land and water management practices which are diagrammatically in Fig. 1.4. All aspects of ecosystems, including their biodiversity, structure, functionality, and benefits to human society, are at risk from these threats. Understanding and addressing the threats to ecosystems is crucial for the conservation of natural resources and the effective administration of ecosystems over the long term. The long-term resilience and well-being of natural systems and human societies can be ensured by the identification and mitigation of these risks, which will also help preserve biodiversity (Middleton 2012).

Wetlands face various threats that can have significant negative impacts on their health, functionality, and overall ecosystem services. Some common threats to wetlands include:

Habitat Loss and Degradation: One of the greatest threats to wetland ecosystems is the destruction and degradation of habitats brought about by human activity.

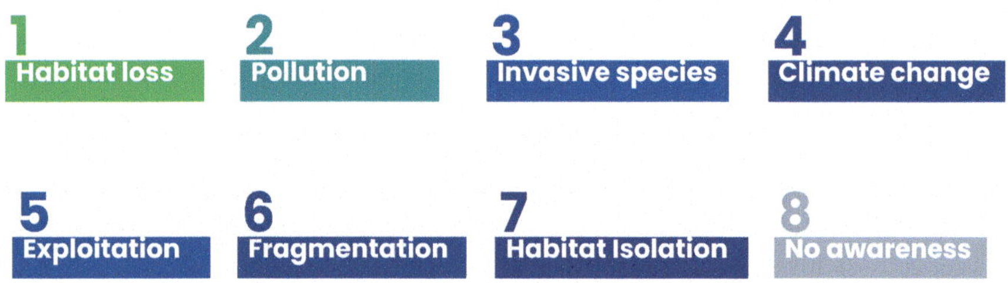

Fig. 1.4 Threats to wetlands

They are often filled up or drained so that the land can be used for farming, building cities, constructing roads, and operating factories. These alterations destroy vegetation and have a negative impact on wetland hydrology, ultimately reducing the effectiveness of the ecosystem as a whole.

Pollution: Industrial discharges, agricultural runoff, sewage, and stormwater runoff from cities all pose threats to wetland water quality. Pollutants such as excess nutrients (eutrophication), heavy metals, pesticides, and pathogens can cause water bodies to become contaminated and lead to a decline in water quality. Pollution can have negative effects on wetland ecosystems, including the plants, animals, and biological processes that live there.

Invasive Species: Invasions or spread of non-native plant and animal species pose a threat to wetland ecosystems. Invasive species have the potential to reduce biodiversity, alter the hydrology of wetland habitats, and disturb ecological processes by outcompeting native species. They may also affect food webs and the way wetland ecosystems function as a whole.

Climate Change: The effects of climate change, which include rising temperatures, altered precipitation patterns, rising sea levels, and more extreme weather, are particularly hard for wetland ecosystems. These shifts have the ability to affect the hydrology of wetland ecosystems, kick-off shifts in vegetation communities, increase salinity levels in coastal wetland areas, and throw off the ecosystem's delicate balance.

Overexploitation: The ecological integrity of wetlands could be jeopardised by the extraction of natural resources in a way that is not sustainable. Overfishing, hunting, logging, and excessive harvesting of wetland flora can all have a harmful effect on wetland species as well as the communities that rely on them. These actions can reduce populations and disrupt ecosystem stability.

Fragmentation and Isolation: Wetlands can become separated and fragmented when man-made features like dams, highways, and farmland are built in their place. Reduced biodiversity and altered ecosystem dynamics are the results of fragmentation, which disrupts ecological connection through its effects on gene flow and species migration.

Lack of Awareness and Understanding: Wetland ecosystems may have degraded in part because people don't realise or don't understand their importance and worth. Inadequate conservation and management efforts, as well as unsustainable land-use practices, are more likely to occur when people lack the necessary information about wetland ecosystems (Mitsch and Gosselink 2015; Van der Valk 2012; Verhoeven et al. 2006).

Coordinated efforts on several fronts are required to counter these threats. Efficient conservation and management practices, the adoption of sustainable land-use practices, strengthened environmental rules and enforcement, promotion of ecosystem restoration, and facilitation of international collaboration for the protection of common ecosystems are all necessary initiatives. To guarantee the long-term health of humans and the world as a whole, it is also important to recognise the interconnectedness of ecosystems the services they provide, and the relevance of protecting ecological integrity and biodiversity. Ecosystems face a wide range of threats, most of which are the result of human actions. Loss of habitat, pollution, climate change, invasive species, overexploitation, fragmentation of habitat, genetic erosion, and a lack of understanding are only a few of the causes of the degradation and loss of ecosystems and the services they provide. Some of the examples are given in Table 1.7.

Table 1.7 Examples of threats wetlands are experiencing

S. no.	Name of wetland	Country of location	Type of wetland	Threats experiencing
1	Everglades	USA	Freshwater Wetland	Habitat loss, water pollution, invasive species, sea-level rise
2	Sundarbans	Bangladesh, India	Mangrove Wetland	Sea-level rise, cyclones, habitat loss, overexploitation of resources
3	Okavango Delta	Botswana	Inland Delta Wetland	Climate change, habitat loss, water extraction, poaching
4	Pantanal	Brazil, Bolivia	Flooded Grassland	Deforestation, agriculture, water pollution, climate change
5	Camargue	France	Salt Marsh Wetland	Coastal development, pollution, invasive species, sea-level rise
6	Kakadu National Park	Australia	Freshwater Wetland	Invasive species, tourism impact, climate change, habitat loss
7	Ciénaga Grande de Santa Marta	Colombia	Coastal Wetland	Pollution, agriculture, overfishing, habitat degradation
8	Danube Delta	Romania, Ukraine	River Delta Wetland	Water pollution, habitat loss, dam construction, agriculture
9	Okefenokee Swamp	USA	Freshwater Wetland	Drought, habitat fragmentation, water pollution, invasive species
10	Wadden Sea	Germany, Netherlands	Coastal Wetland	Climate change, pollution, habitat degradation, sea-level rise
11	Hula Lake	Israel	Freshwater Wetland	Water diversion, habitat loss, invasive species, pollution
12	Tonle Sap Lake	Cambodia	Seasonal Floodplain	Overfishing, habitat loss, water pollution, dam construction
13	Lake Baikal	Russia	Freshwater Wetland	Pollution, invasive species, climate change, habitat degradation
14	Térraba-Sierpe National Wetlands	Costa Rica	Mangrove Wetland	Pollution, habitat loss, aquaculture, overfishing
15	Chilika Lake	India	Brackish Water Lagoon	Siltation, overfishing, habitat loss, pollution
16	Tana River Delta	Kenya	River Delta Wetland	Deforestation, agriculture, water pollution, habitat loss
17	Lagoa do Peixe	Brazil	Coastal Lagoon	Pollution, habitat degradation, water diversion, climate change
18	Lake Titicaca	Peru, Bolivia	Freshwater Lake	Pollution, habitat degradation, overfishing, climate change
19	Rann of Kutch	India, Pakistan	Desert Salt Marsh	Pollution, habitat loss, overgrazing, salt mining
20	Chokpak Wetland	Kazakhstan	Saline Wetland	Pollution, habitat degradation, water diversion, climate change

References

Anderson JT, Davis CA (eds) (2013) Wetland techniques: volume 1: foundations, vol 1. Springer Science & Business Media

Batzer DP, Baldwin AH (eds) (2012) Wetland habitats of North America: ecology and conservation concerns. Univ of California Press

Batzer DP, Sharitz RR (eds) (2014) Ecology of freshwater and estuarine wetlands. Univ of California Press

Dennison MS, Berry JF (1993) Wetlands: guide to science, law and technology. Cambridge University Press

Denyer K, Peters M (2022) Life in the Shallows: The Wetlands of Aotearoa New Zealand. Massey University Press

Finlayson C, Everard M, Irvine K, McInnes RJ, Middleton BA, van Dam A, Davidson N (2018a) The wetland book I: Structure and function, management and methods. Springer

Finlayson C, Milton GR, Prentice RC, Davidson N (2018b) The wetland book II: distribution, description and conservation. Springer

Gearey M, Church A, Ravenscroft N (2020) English Wetlands: Spaces of nature, culture, imagination. Springer Nature

Giblett R (2019) Postmodern wetlands: Culture, history, ecology. Edinburgh University Press

Kar D (2013) Wetlands and Lakes of the World. Springer, New Delhi, p 687

Keddy PA (2010) Wetland ecology: principles and conservation. Cambridge university press

Kent DM (ed) (2000) Applied wetlands science and technology. CRC Press

Křeček J, Krecek J, Haigh M (eds) (2006) Environmental role of wetlands in headwaters, vol 63. Springer Science & Business Media

Marble AD (1991) A guide to wetland functional design. CRC Press

Middleton BA (ed) (2012) Global change and the function and distribution of wetlands, vol 1. Springer Science & Business Media

Mitsch WJ, Gosselink JG (2015) Wetlands. John Wiley & Sons

Van der Valk A (2012) The biology of freshwater wetlands. Oxford University Press

Verhoeven JT, Beltman B, Bobbink R, Whigham DF (eds) (2006) Wetlands and natural resource management, vol 190. Springer Science & Business Media

Saline Wetland Ecosystems 2

Abstract

This chapter of this book, titled "Saline Wetland", delves into a specialised subset of wetland ecosystems characterised by their high salinity levels. Often overlooked, saline wetlands possess distinct ecological roles and challenges, warranting focused exploration. The chapter is structured into various sections, each illuminating different facets of saline wetlands. It commences by elucidating the stark contrast between saline and freshwater wetlands, emphasising the critical importance of recognising these differences for effective management. Amid the climate crisis, saline wetlands emerge as crucial players in mitigating global warming, serving as potent carbon sinks, a topic thoroughly examined in a subsequent section. Classification of saline wetlands based on salinity, geography, and vegetation provides deeper insights into their diversity and functions within the broader wetland ecosystem. The discussion extends to the ecosystem services rendered by saline wetlands, encompassing habitat provision, water purification, and economic benefits. However, these valuable ecosystems face a myriad of threats, including habitat loss, pollution, invasive species, and the impacts of climate change, all dissected in detail. This chapter offers readers a comprehensive understanding of saline wetland ecosystems, their significance in climate change mitigation, and the imperative of conservation efforts in safeguarding these unique environments.

Keywords

Ecosystem Services · Global Warming · Saline Wetlands · Threats

L. K. Sharma, R. Naik, *Conservation of Saline Wetland Ecosystems*,
https://doi.org/10.1007/978-981-97-5069-6_2

Saline wetlands are a special and varied form of wetland ecosystem. They are found in the locations with high salt content, such as the shoreline, salt flats, and salt pans. Salty or brackish water is a defining feature of these types of wetlands, which might be the result of tidal inflows, saltwater intrusion, or natural geological processes. Saline wetlands are comprised of variety of ecosystems, including salt marshes, mangrove forests, seagrass beds, and brackish lagoons. Biological processes and species compositions in each of these environments are distinct from one another. Their importance is given in Fig. 2.1.

One of the key reasons for the great ecological significance of saline wetlands is that they can host a wide variety of plant and animal species, each of which is uniquely adapted to live in these challenging conditions. Many species of migratory birds, fish, and other creatures rely heavily on these ecosystems for reproduction, nesting, and foraging. Salt-tolerant grasses, mangroves, and halophytic vegetation are just a few of the examples of plants that may thrive in these environments. It's possible that these plants do well in extremely salty conditions.

High rates of productivity and nutrient cycling are two of the most distinguishing features of saline wetland ecosystems. Despite the challenges, these wetland ecosystems support a rich food web and add significantly to the health of the larger biome. They are crucial nurseries for commercially important fish species and recreationally important fish species because they provide a location for juvenile fish and invertebrates to feed and shelter. Saline wetlands help maintain water quality in another way: by filtering out sediments, fertilisers, and toxins, they make the water cleaner and clearer.

In addition to their role in climate regulation, saline wetland habitats play a crucial role in the defence of vulnerable coastal regions. They serve as natural barriers to reduce the destructive potential of wave energy during extreme weather events like storm surges and coastal erosion. In particular, mangroves help stabilise shorelines and protect against erosion because of their enormous root systems. In addition, these wetland habitats store large quantities of carbon, which mitigates the results of global warming. With greenhouse gas emissions on the rise and global

Fig. 2.1 Importance of saline wetlands

warming speeding up as a result, these wetlands' ability to absorb and store carbon is of critical importance (Chauhan and Gopal 2016).

The environmental, cultural, social, and economic value of saline wetland regions cannot be overstated. For millennia, indigenous groups have developed unique cultural practices and traditional knowledge related to these ecosystems as a result of their proximity to and interaction with them. Saline wetland habitats attract tourists and boost local economies because of the abundance of options for outdoor recreation, such as birdwatching, boating, and fishing. People who are interested in nature and ecotourism often visit these locations because of the stunning natural beauty, abundance of wildlife, and tranquillity they offer.

Despite their importance, saline wetland habitats face numerous threats and challenges. Coastal development, land reclamation, pollution, overfishing, and climate change all pose serious threats to the health of these ecosystems and their ability to function normally. Loss of biodiversity and the services given by ecosystems can be the result of human-caused changes in hydrological patterns, salt levels, and the quality of habitat.

If saline wetland regions are to survive and continue providing the services they provide now and, in the future, they must be protected and maintained in a sustainable way. Many efforts are under way to preserve and restore these priceless ecosystems, including the establishment of protected areas, the adoption of integrated coastal zone management, and the promotion of sustainable fishing and land-use practices. Collaborative approaches involving scientific communities, local communities, NGOs, and government agencies are necessary for successful conservation projects. Some of the examples are given in Table 2.1.

Table 2.1 Examples of saline wetlands

S.no.	Name of saline wetland	Country of location	Importance	Threats experiencing
1	Chott el Jerid	Tunisia	Ecological, Cultural, Economic	Water extraction, habitat degradation, climate change
2	Sambhar Lake	India	Ecological, Cultural, Economic	Salt extraction, pollution, habitat loss, water diversion
3	Great Salt Lake	USA	Ecological, Recreational	Water diversion, pollution, habitat degradation, invasive species
4	Lake Urmia	Iran	Ecological, Cultural, Economic	Water diversion, pollution, habitat degradation, climate change
5	Maranjab Salt Desert	Iran	Ecological, Recreational	Overgrazing, habitat degradation, invasive species
6	Etosha Pan	Namibia	Ecological, Wildlife Habitat	Drought, pollution, habitat degradation, climate change
7	Lake Natron	Tanzania	Ecological, Wildlife Habitat	Water diversion, pollution, habitat degradation, invasive species

(continued)

Table 2.1 (continued)

S.no.	Name of saline wetland	Country of location	Importance	Threats experiencing
8	Salar de Uyuni	Bolivia	Ecological, Recreational	Mining, tourism impact, pollution, habitat degradation
9	Lake Eyre	Australia	Ecological, Cultural	Drought, water diversion, habitat degradation, pollution
10	Lop Nur	China	Ecological, Cultural	Water diversion, habitat degradation, pollution, climate change
11	Badwater Basin	USA	Ecological, Recreational	Mining, tourism impact, pollution, invasive species
12	Lake Abert	USA	Ecological, Recreational	Water diversion, pollution, habitat degradation, climate change
13	Lake Tuz	Turkey	Economic, Ecological	Salt mining, pollution, habitat degradation, water diversion
14	Koyashskoye Salt Lake	Russia	Ecological, Cultural	Pollution, habitat degradation, water diversion, invasive species
15	Sebkha Sejoumi	Tunisia	Ecological, Cultural	Habitat degradation, pollution, water extraction, climate change
16	Lake Tengiz	Kazakhstan	Ecological, Wildlife Habitat	Pollution, water diversion, habitat degradation, invasive species
17	Lake Qarun	Egypt	Economic, Ecological	Pollution, habitat degradation, water extraction, climate change
18	Kara-Bogaz-Gol	Turkmenistan	Economic, Ecological	Water diversion, habitat degradation, pollution, invasive species
19	Soda Lakes of Kenya	Kenya	Ecological, Cultural	Habitat degradation, water extraction, pollution, climate change
20	Karum Salt Lake	Ethiopia	Economic, Ecological	Salt mining, pollution, habitat degradation, water diversion

2.1 Saline Wetland Characteristics

Wetlands can be either saltwater (or "saline") or freshwater (or "fresh") in composition. Plant and animal adaptations, as well as biological activities, help differentiate these two types of wetlands. The following is a summary of key differences between saline and freshwater wetland types which are also diagrammatically represented in Fig. 2.2.

	Fresh water bodies	Saline water bodies
Salinity	Less than 1 gm/l	Between 3 gm/l-300 gm/l
Density	Low density	High density
Mineral composition	Low	High
Examples	Ponds, streams, rivers	**Inland & Coastal**

Fig. 2.2 Comparison between saline and freshwater bodies

Water Salinity: One key distinction between freshwater and saltwater wetlands is the salinity of the water. In saline wetland environments, the water's salinity often rises above that of the open sea. In contrast, low salinity or fresher qualities of water are typical with freshwater wetlands.

Plant Adaptations: Plants that have adapted to thrive in high-salt environments are called halophytes. Salt marshes are home to these plants. Salt glands and succulent leaves are just two examples of how these plants have adapted to survive in salty environments. The emergent plants of freshwater wetlands, such as cattails, reeds, and sedges, are characterised by their adaptation to the conditions of freshwater.

Animal Adaptations: Crustaceans, fish with specialised diets, and migrating birds are just some of the salt-resistant animals that make their homes in saline wetland settings. Some animals have adapted to the high salinity of the ocean by changing their behaviour to avoid it, while others have adapted physiologically to live in the water. However, many different kinds of animals and plants have adapted to live in freshwater wetlands because of the habitats they provide. Amphibians, ducks, and freshwater fish are all examples of this group.

Species Diversity: In general, there is less species diversity in saline wetlands than in freshwater wetlands because of the harsh circumstances induced by high salinity. This is because it is more challenging to live in saline wetlands. Despite this, they have managed to keep a diverse array of species that are unique and adapted to their environment. Because of the relative lack of adversity present in freshwater wetlands, they typically support a richer flora and fauna than their saltwater counterparts.

Ecological Functions: Wetlands, whether saltwater or freshwater, have different biological roles in their surrounding ecosystems. Saline wetlands are crucial to the process of salt and mineral cycling due to the high salt concentrations they carry. In addition to trapping sediments and contaminants, these wetlands filter them out naturally. They also encourage the development of specialised plants that are resistant to high amounts of salt, and they provide a haven for nesting, breeding, and foraging by migratory birds. Freshwater wetlands have a greater effect on hydrological processes such as water storage, flood mitigation, and water purification because of their location. They also play an important role in the nitrogen cycle, provide homes for a diverse range of freshwater creatures, and help keep the water clean.

Threats and Conservation: Wetland ecosystems, whether freshwater or marine, face their own set of difficulties that require tailored conservation efforts. The loss of land, the entry of pollutants, and the diversion of water are all threats to the saline wetland's delicate salt and water dynamics. Human activities including urbanisation and agricultural operations pose threats to freshwater wetland ecosystems in the form of habitat loss, water supply contamination, and disturbance of natural hydrological cycles. The protection of freshwater and marine wetland species depends critically on the maintenance and restoration of their respective habitats. The salinity of the water, the adaptations of plants and animals to the environment, the diversity of species, and the ecological roles played by freshwater and saltwater wetlands all distinguish them as distinct ecosystems. Understanding these differences is crucial for the conservation and management of these important wetland ecosystems (Hammer 1986).

2.2 Importance of Saline Wetlands for Reducing Global Warming

Saline wetland ecosystems are particularly important in the context of global warming and climate change. These unique ecosystems contribute to mitigating the consequences of climate change in a number of ways, despite being commonly associated with high salinity and challenging circumstances. Some of the ways in which saline wetlands exacerbate the effects of global warming are as follows:

Carbon Sequestration: Carbon dioxide is captured from the atmosphere and sequestered in remarkably efficient ways by saline wetland ecosystems, such as salt marshes and mangrove forests. Carbon is captured from the air and stored in the abundant flora and organic matter typical of wetland habitats. Plants take in carbon dioxide from the air during photosynthesis and use it to fuel their growth; the excess is stored as biomass in the soil. The slow rate of decomposition of organic matter caused by the high salinity of these wetland habitats results in the long-term storage of carbon. Preserving and rebuilding saline wetland ecosystems can boost their ability to sequester carbon. The atmospheric concentration of greenhouse gases will decrease as a result of this.

Methane Emission Regulation: Saline wetland ecosystems, especially those that are waterlogged or anaerobic, have the potential to produce methane (CH_4), one of the most potent greenhouse gases. But they can also act as methane sinks when the right conditions are met. Sulphate in salt marsh sediments can encourage the development of sulphate-reducing bacteria, which compete with methane-producing bacteria for food sources. This rivalry causes organic carbon to be converted into less harmful forms, which in turn reduces methane generation and benefits the environment. This means that salty wetland methane emissions may be lower than those of freshwater wetlands, aiding in the fight against global warming.

Coastal Protection: Salt marshes and mangrove forests are just two examples of saline wetland habitats that protect coasts from the damaging effects of climate change, such as increasing sea levels and storm surges. When natural barriers like these wetland areas are available, incoming waves and currents lose energy and move more slowly. Dense vegetation and complex root systems are shared characteristics of salt marshes and mangroves, both of which aid in protecting shorelines and preventing erosion by stabilising sediments, dissipating wave energy, and trapping material. Because of their ability to shield coastal areas, saline wetlands serve a significant role in limiting the amount of damage caused by storms and maintaining the communities and infrastructure that are located nearby.

Water Quality Improvement: Because of this, saline wetlands are able to purify the water that passes through them. They help trap sediments, fertilisers, and toxins, preventing them from accessing water bodies further downstream. By removing contaminants and surplus nutrients like nitrogen and phosphorus, this filtering technique improves water quality by reducing eutrophication and the risk of harmful algal blooms. Increased resistance to climate change is facilitated by saline wetland habitats, which not only increase water quality but also preserve the health and functionality of aquatic ecosystems.

Biodiversity Conservation: Rare and specialised species that have developed the ability to live and even thrive in salty marsh environments are found there. Fish, invertebrates, migrating birds, and salt-tolerant plants are just some of the many species that call these habitats home. It is crucial to ensure the survival of saline wetland ecosystems and the animals that live there. The continued resilience and adaptability of ecosystems in the face of changing climatic conditions depends on the protection of biodiversity (Weller 1999).

2.3 Classification of Saline Wetlands

These classifications are not exclusive of one another and that some wetland types, depending on their location, hydrology, and neighbouring ecosystems, may display characteristics that are diagnostic of more than one kind. Saline wetland classifications which are diagrammatically represented in Fig. 2.3 aid in learning about the unique ecological features of these places and guiding conservation and management efforts to protect these vital ecosystems.

2.3.1 Classification Is Based on Distance from Ocean

They are of two types, coastal saline wetlands and inland saline wetlands. They are as follows:

Fig. 2.3 Classification of saline wetlands

Coastal wetlands: Coastal wetlands are a highly significant and dynamic habitat found at the edge of land and sea. Wetlands like these are common along the coasts of every continent, and they play a crucial role in coastal ecosystems by providing habitat, shielding coastlines, and sustaining a wide range of plant and animal life. Coastal wetlands encompass a wide range of ecosystems, from salt marshes and mangrove forests to coastal lagoons, each with its own unique characteristics and biological roles.

Inland saline wetlands: Unique and fascinating, inland saline wetland ecosystems are defined by high salt concentrations in the water and soil of the wetland ecosystems in which they are located. These wetlands play important ecological roles in the ecosystems they inhabit and can be found in a wide range of topographic settings, from salt flats and salt pans to saline lakes. There is great scientific interest in inland saline wetland habitats because of their ecological significance and the crucial ecosystem services they provide. These wetland habitats are home to a surprising diversity of plant and animal life, including specialised microorganisms, invertebrates, and plants, despite their environments' austere appearance. Some of these species actually thrive in salty environments, while many others have evolved mechanisms that allow them to survive in such conditions. The high salinity acts as a barrier, limiting the number of competitors and predators in the area, which in turn promotes the development of specialised adaptations. Inland saline wetland ecosystems also play a significant role in water and nutrient cycle. They act as natural filters by collecting sediments and storing excess nutrients, so limiting their transport into downstream water sources. The detrimental impacts of nutrient contamination on aquatic ecosystems can be mitigated because of the work done by these wetland areas. Salty inland marshes can also have a major effect on the economy. They probably hold significant mineral reserves that may be mined for profit, such as salt, gypsum, or borates. The availability of these assets improves the economic outlook for the regions surrounding them.

2.3.2 Classification of Saline Wetlands as per Chemical Property

The salinity of the water is one chemical characteristic that can be used to categorise saline wetlands. It's crucial to remember that marsh salinity can change with the seasons or because of external causes like freshwater runoff, precipitation, and tidal changes. Understanding how plants and animals have adapted to varying salinity levels is aided by this chemical classification. The term "salinity" is used to describe the amount of dissolved salts in water, most commonly sodium chloride. Several different kinds of saline wetlands can be distinguished by their salt levels.

Freshwater Wetlands: Freshwater wetland ecosystems are distinguished by their low salinity, often below 0.5 Parts Per Thousand (ppt). Rivers, streams, and even rainwater have a significant impact on them because they are freshwater sources. Marshes, swamps, and peatlands are all examples of freshwater wetlands, and they support a rich flora that is adapted to low-salinity conditions.

Brackish Wetlands: The salinity of brackish wetlands typically ranges from about 0.5 to 30 ppt. They develop in the areas where freshwater and saltwater meet, such as estuaries, tidal marshes, and coastal lagoons. Tidal effects on brackish wetlands result in very variable salinities, making this type of environment suitable for a wide variety of freshwater and marine organisms.

Saline Wetlands: Saline wetlands often contain salt concentrations of 30 ppt or thousand or higher. Salt marshes form when the rate of evaporation exceeds the rate of freshwater absorption, as it does in arid and coastal areas. Wetland environments that are rich in salt include salt marshes, mangrove swamps, saline ponds, and saltpans. Halophytic vegetation, which thrives in extremely salty environments, finds a natural home there.

Hypersaline Wetlands: When the salinity of a wetland is extremely high, called hypersaline, often above 50 ppt, it can become completely saturated. Salt flats, salt pans, and some inland saline lakes are examples of this type of wetland since they all have very high rates of evaporation yet get very little freshwater flow. Vegetation is rarely found in hypersaline wetlands, and the unique chemical conditions found there have allowed for the evolution of specialised microbial communities.

2.3.3 Classification of Saline Wetlands as per Geological Property

There are many different types of saline wetlands, each distinguished by their own unique combination of geological features and formation processes. These groups help explain the diversity of saline wetland ecosystems by highlighting their individual characteristics and histories. These groups provide a foundation for learning about the great range of saline wetland habitats and their genesis in the earth's history. It's important to remember, nevertheless, that wetland ecosystems are dynamic, ever-changing places, where features and qualities may overlap. The unique features of a saline wetland may also be affected by regional climate, hydrology, and other environmental factors. Saline wetlands can be broken down into numerous subgroups based on their individual geological properties and formation processes. Their unique features can be influenced by a number of environmental factors, including the local climate, hydrology, and others. The following are examples of common classifications for saline wetlands:

Sabkha Wetlands: Sabkhas, which can be found both on the coast and inland, are characterised by their salt crusts and salt pans. They thrive in places with low precipitation and significant evaporation, like the dry, arid regions of the world. Sabkha wetlands are formed when salt minerals and sediments build up, making available unique habitats for salt-adapted plant species.

Saline Karst Wetlands: There are soluble rock formations like limestone and gypsum present in these wetland areas, making these regions karst landscapes. Saline karst wetland ecosystems are characterised by the presence of an excess of dissolved salts and can be found in areas where underground water emerges, usually in the form of springs or sinkholes. Inside these wetlands, saline springs, sinkholes, and the ecosystems that depend on them might be found.

Saline Playa Wetlands: It is common to find playas, or transient sources of water, in arid and semiarid environments. Shallow depressions that fill with water sometimes and become saline due to evaporation and mineral concentration characterise saline playa wetlands. Saline playa wetlands are common in dry climates. Desert basins are characterised by their ability to shelter unique flora and fauna that have evolved to survive in the arid environment.

2.3.4 Classification of Saline Wetlands as per Biological Property

Considerations for plant and animal adaptations to high-salinity conditions should be made when categorising saline wetland regions based on their biological traits. There is a close relationship between the chemical make-up and physical features of saline wetland ecosystems and their biological classification. It is crucial for the preservation and management of these wetlands to have an understanding of the biological factors that shape their structure and function. The form and function of these wetland habitats are determined by the species' adaptations to high salinity levels. The most common categories are as follows:

Salt Marshes: One of the distinguishing features of salt marshes is the presence of salt-resistant plants, known as halophytes. Different adaptations, such as the ability to expel salt or store it in tissues that are specifically intended for the purpose, allow these plants to flourish in salty environments. Numerous species of birds, fish, and invertebrates rely on salt marshes for survival. They also provide vital feeding and breeding grounds for species that migrate.

Mangrove Forests: Mangroves are a type of tree or shrub that thrives in coastal areas with high levels of salt. These unique ecosystems are restricted to the tropical and subtropical coasts, and their unique characteristics allow them to thrive in extremely salty water. Fish and other marine animals rely heavily on the mangrove nurseries. Mangroves also prevent erosion of the coast and act as vital carbon sinks.

Seagrass Meadows: Communities of blooming plants that have adapted to flourish in saline circumstances while remaining entirely submerged form the seagrass meadows, an underwater environment. They help stabilise sediments, supply habitat for marine species, and keep food webs diversified, and they can be found in the shallow coastal areas and estuaries. They populate both marine and freshwater ecosystems.

Halophytic Wetlands: Wetlands that are dominated by halophytic plants, or those that have adapted to grow well in high salt environments, are called halophytic wetlands. The various forms of wetland ecosystems include salt pans, salt flats, and saline ponds. In spite of their limited plant diversity, halophytic wetland ecosystems are home to specialised communities of salt-tolerant plants and the animals that rely on them.

Hypersaline Ponds and Lakes: Most aquatic organisms cannot survive in the high quantities of salt seen in hypersaline ponds and lakes. This makes maintaining life difficult. The bulk of known life forms cannot tolerate the high salt content of these settings, although they do provide a home for unique populations of halophilic microorganisms and extremophiles (Williams 1996).

2.4 Ecosystem Services of Saline Wetlands

Unique to wetlands with high salt levels, saline wetland ecosystem services are crucial to the health of the ecosystem and human well-being. Some of the most significant reasons why saline wetland ecosystems are so important for the ecosystem services they give are represented diagrammatically in Fig. 2.4.

Fig. 2.4 Importance of saline wetland ecosystem services

Biodiversity Support: Rare and specialised species abound in saline wetlands, many of which have adapted to the high levels of salt in their natural habitat. Numerous species of plants, animals, and microbes have adapted to life in salty marsh areas and make use of these ecosystems. Numerous plant and animal species, including some that can be found nowhere else on Earth, find refuge in saline wetland ecosystems. This aids the overall health of ecosystems and the variety of life within them.

Habitat Provision: Migratory birds, waterfowl, and many different kinds of aquatic organisms all rely on saline wetland areas as critical habitats on their long journeys. These wetlands are used by both permanent residents and seasonal migrants as habitats, providing food, water, and shelter for a wide variety of animals. In order to maintain these ecosystems and sustain the life cycles and population dynamics of a wide range of species, it is essential to preserve and safeguard saline wetland habitats.

Water Regulation and Filtration: Saline wetland ecosystems aid in water level regulation because they act as natural reservoirs and water filtering and purification systems. They aid in the control of water flow, mitigation of flooding, and groundwater replenishment. They can act as natural filters by removing harmful substances from the water and absorbing excess nutrients and sediments.

Carbon Sequestration: Saline wetland habitats aid in the mitigation of climate change by sequestering carbon. Although saline wetlands' rates of carbon sequestration are often lower than those of other types of wetlands, they do play a role in the process. In order to reduce the amount of greenhouse gases in the atmosphere, saline wetlands rely on the plants and organic matter found there to capture and store carbon.

Erosion Control: Because of their potential to stabilise shorelines and prevent land deterioration, saline wetland regions are useful in the prevention and management of soil erosion. They prevent soil erosion along coasts and around the edges of saltwater bodies by providing structural stability to the soil through their foliage and root systems.

Cultural and Traditional Values: Salt marshes have great symbolic and historical meaning for many people. They are often intertwined with the customs and religious practices of the local community. As places where people congregate for rituals and other cultural events, saline wetlands may have considerable cultural significance. It is crucial that these cultural values be kept and cherished for the sake of cultural history and the general well-being of the community.

Education and Research: Saline wetlands provide opportunities for education, research, and scientific study. They offer unique environments for studying the adaptations and ecological processes specific to saline conditions. Saline wetlands contribute to scientific knowledge, enhancing our understanding of specialised ecosystems, biodiversity, and ecosystem functioning.

Economic Benefits: Saline wetlands can generate economic benefits through activities such as tourism, recreational use, and traditional livelihoods. These wetlands attract visitors interested in their unique flora, fauna, and landscapes, contributing to nature-based tourism and local economies. Saline wetlands also support traditional livelihoods, such as fishing and salt harvesting, providing economic opportunities for local communities.

2.4.1 Types of Saline Wetland Ecosystem Services

Provisioning Ecosystem: Saline wetlands provide several provisioning ecosystem services that directly benefit human populations, some of which are mentioned in Table 2.2. Here are some examples:

Salt Production: Saline marshes are known for their high salt content, making them an ideal area for salt production. Salt has been mined from these wetlands for thousands of years; it has numerous applications in the home, the workplace, and the medical field. In addition to helping the local economy, the harvesting of salt from salt marshes also provides locals with a source of income.

Fisheries: The high salinity of some salt marshes supports fish populations of both native and invasive species. Due to their importance in fish reproduction, these wetlands are a boon to local economies and provide residents with food and employment. Through the use of sustainable fishing methods in salt marshes, fishermen may protect their livelihoods and increase food security.

Aquaculture: Saline wetland habitats are suitable for aquaculture activities, particularly for species that can survive in high salt conditions. In some types of salt marshes, shrimp farming has become an important economic and job driver. Sustainable aquaculture that has minimal negative effects on the environment can be achieved through the use of managed practices.

Grazing and Livestock Watering: The creation of grazing grounds for animals is possible in salty wetland areas with brackish water or salt-tolerant vegetation. The cattle industry benefits from the year-round or seasonal usage of these wetland areas as grazing pastures due to the increased availability of water for the animals.

Traditional and Medicinal Plant Harvesting: Plants in saline marshes often have unusual adaptations that allow them to thrive despite the high salt levels. Some of these plants are picked for their useful qualities, and they are also used in alternative medicine and traditional medicine. Local communities can benefit from harvesting these plants because they can be used in food preparation, traditional medicine, and decorative purposes.

Reed and Phragmites Harvesting: Dense stands of reeds or phragmites in salt marshes can be harvested for a number of raw materials. Reeds can be used in numerous crafts, such as thatching roofs, making baskets and mats, and many others. These wetlands contribute to the local economies through the responsible harvest of its resources.

Traditional Hunting and Gathering: Traditional human activities such as hunting and gathering might be practised in salt marshes. Communities in the vicinity of wetlands may rely on these resources for a wide range of purposes, including food gathering, waterfowl hunting, and the gathering of raw materials for traditional crafts (such as reeds, fibres, or shells).

Table 2.2 Examples of saline wetlands for provisioning services

S.no.	Name of saline wetland	Country of location	Type of saline wetland	Provisioning services ecosystem services provided
1	Chott el Jerid	Tunisia	Salt Flat	Salt extraction, mineral resources
2	Sambhar Lake	India	Salt Lake	Salt extraction, brine shrimp harvesting, salt-based products
3	Great Salt Lake	USA	Terminal Lake	Salt production, brine shrimp harvest, mineral extraction
4	Lake Urmia	Iran	Endorheic Lake	Fisheries, traditional salt harvesting
5	Maranjab Salt Desert	Iran	Playa	Salt collection, mineral resources, local livelihoods
6	Etosha Pan	Namibia	Salt Pan	Salt collection, wildlife grazing
7	Lake Natron	Tanzania	Alkaline Lake	Salt production, brine shrimp harvesting, traditional uses
8	Salar de Uyuni	Bolivia	Salt Flat	Salt extraction, lithium mining, tourism, mineral resources
9	Lake Eyre	Australia	Salt Pan	Salt harvesting, brine shrimp collection, local uses
10	Lop Nur	China	Salt Pan	Salt production, traditional uses, historical significance
11	Badwater Basin	USA	Salt Pan	Salt collection, unique geological features, tourism
12	Lake Abert	USA	Endorheic Lake	Salt extraction, mineral resources
13	Lake Tuz	Turkey	Salt Lake	Salt production, mineral resources
14	Koyashskoye Salt Lake	Russia	Salt Lake	Salt harvesting, mineral resources, local livelihoods
15	Sebkha Sejoumi	Tunisia	Salt Pan	Salt extraction, salt-based products, local uses
16	Lake Tengiz	Kazakhstan	Salt Lake	Salt extraction, fishing, traditional uses
17	Lake Qarun	Egypt	Salt Lake	Salt production, fisheries, salt-based products
18	Kara-Bogaz-Gol	Turkmenistan	Lagoon	Salt collection, fishing, mineral resources, local livelihoods
19	Soda Lakes of Kenya	Kenya	Saline Lakes	Salt extraction and traditional salt production, local uses
20	Lake Assal	Djibouti	Crater Lake	Salt extraction and traditional salt production, mineral resources
21	Oroumieh Lake	Iran	Endorheic Lake	Fisheries, salt production, local livelihoods
22	Salinas Grandes	Argentina	Salt Flat	Salt extraction, lithium mining, local economy
23	Lake Uyuni	Bolivia	Salt Flat	Salt extraction, lithium mining, tourism, mineral resources
24	Lake MacLeod	Australia	Salt Lake	Salt production, mineral resources, local economy

Tourism and Recreation: The unique landscapes, diverse wildlife, and cultural value of saline wetlands make them attractive destinations for tourists and nature lovers. Community economies may profit from tourism and recreation centred on wetland regions like birdwatching, wildlife photography, boating, and hiking nature paths. Hospitality, guiding, and other tourism-related businesses stand to gain from these incentives.

2.4.2 Regulating Ecosystem Services

Saline wetlands help maintain ecological balance and provide a variety of ecosystem services that govern the natural environment, making them useful to both natural ecosystems and human communities which are mentioned in Table 2.3. Some of the regulation ecological services provided by salt marshes are as follows:

Water Regulation: The negative effects of droughts and floods can be lessened because of the role saline wetland regions play as natural controllers of water flow. The planet is dotted with saline wetland zones. They may absorb and store more water, reducing the likelihood of flooding further downstream during periods of heavy precipitation. They are able to save water because of this feature. By releasing water that has been retained during drought, wetlands help to regulate water availability and maintain steady water levels in the areas surrounding them. This will allow for more stable water levels in the neighbouring communities.

Water Purification: By filtering out sediments, pollutants, and excess nutrients, saline wetlands play a crucial role in the water purification process. They serve as a form of natural purification by collecting and decomposing harmful substances found in water. In general, this improves the water's quality. Wetlands can reduce the amount of pollutants and nutrients that reach downstream water sources. This safeguards both our freshwater supplies and our aquatic ecosystems.

Climate Regulation: Saline wetland ecosystems contribute to climate regulation through activities such as CO_2 sequestration and GHG emission reduction. Wetland areas help in the fight against climate change because their plants and soils are excellent at absorbing and storing carbon. Because of their distinct hydrological and climatic effects on the surrounding area's vegetation as well as the temperature and humidity levels, wetlands can influence regional climate patterns.

Erosion Control: Salty and brackish wetlands are crucial in halting soil erosion. As natural barriers, they assist shorelines in remaining stable and lessen the impact of waves and currents along riverbanks and beaches. Wetlands have a lower rate of erosion and land loss because the plants and root systems there work collectively to bind soil particles together. By reducing erosion rates, wetlands protect neighbouring habitats, infrastructure, and agricultural areas.

Flood Mitigation: In the event of heavy rain or high-water levels, saline wetland areas can act as natural sponges and absorb the excess. Wetlands help mitigate

Table 2.3 Examples of saline wetlands for regulating services

S.no.	Name of saline wetland	Country of location	Type of saline wetland	Regulating services ecosystem services provided
1	Chott el Jerid	Tunisia	Salt Flat	Climate regulation, flood control, erosion prevention
2	Sambhar Lake	India	Salt Lake	Flood control, water purification, erosion prevention
3	Great Salt Lake	USA	Terminal Lake	Climate regulation, flood control, water purification
4	Lake Urmia	Iran	Endorheic Lake	Climate regulation, flood control, water purification
5	Maranjab Salt Desert	Iran	Playa	Erosion prevention, water purification, climate regulation
6	Etosha Pan	Namibia	Salt Pan	Climate regulation, flood control, water purification
7	Lake Natron	Tanzania	Alkaline Lake	Water purification, climate regulation, erosion prevention
8	Salar de Uyuni	Bolivia	Salt Flat	Climate regulation, flood control, erosion prevention
9	Lake Eyre	Australia	Salt Pan	Climate regulation, flood control, water purification
10	Lop Nur	China	Salt Pan	Erosion prevention, climate regulation, flood control
11	Badwater Basin	USA	Salt Pan	Climate regulation, flood control, water purification
12	Lake Abert	USA	Endorheic Lake	Flood control, water purification, climate regulation
13	Lake Tuz	Turkey	Salt Lake	Climate regulation, flood control, erosion prevention
14	Koyashskoye Salt Lake	Russia	Salt Lake	Erosion prevention, climate regulation, flood control
15	Sebkha Sejoumi	Tunisia	Salt Pan	Climate regulation, flood control, erosion prevention
16	Lake Tengiz	Kazakhstan	Salt Lake	Climate regulation, flood control, water purification
17	Lake Qarun	Egypt	Salt Lake	Water purification, erosion prevention, climate regulation
18	Kara-Bogaz-Gol	Turkmenistan	Lagoon	Climate regulation, flood control, erosion prevention
19	Soda Lakes of Kenya	Kenya	Saline Lakes	Flood control, water purification, climate regulation
20	Lake Assal	Djibouti	Crater Lake	Erosion prevention, climate regulation, flood control
21	Oroumieh Lake	Iran	Endorheic Lake	Climate regulation, flood control, water purification
22	Salinas Grandes	Argentina	Salt Flat	Water purification, climate regulation, erosion prevention
23	Lake Uyuni	Bolivia	Salt Flat	Erosion prevention, flood control, climate regulation
24	Lake MacLeod	Australia	Salt Lake	Climate regulation, water purification, flood control

flood damage because they store and slowly release water. They act as floodplains and flood storage regions naturally, lowering flood peaks and lowering the risk of flooding further downstream.

Nutrient Cycling: Salt marshes play a crucial role in nutrient cycling, especially for nitrogen and phosphorus. They prevent an overabundance of nutrients from entering water supplies by ingesting, transforming, and storing them. They aid in nutritional balance by acting in this way. By controlling nutrient levels, we can protect aquatic ecosystems, lessen the risk of eutrophication, and improve the well-being of aquatic plants and animals.

Coastal Protection: Salty wetland areas in strategic locations along coastlines serve as natural buffers against the destructive effects of storm surges and erosion. They dissipate the force of the waves, making them less damaging and protecting coastal communities from flooding and erosion. Protecting significant coastal habitats and the people and infrastructure that depend on them is a primary function of wetland areas.

Wildlife Habitat and Biodiversity Support: Many plant and animal species, especially those that have adapted to high levels of salt, rely on saline wetland habitats for survival. Many different kinds of animals live in and around these wetlands; some of them are permanent residents, while others are migratory birds and ducks. Wetlands are important to the local ecosystem because they provide animals with a place to dwell, food, and a place to reproduce.

2.4.3 Supporting Ecosystem Services

Saline wetlands provide several supporting ecosystem services that are essential for the functioning of ecosystems and the well-being of human communities which are mentioned in Table 2.4. Here are some examples of supporting ecosystem services provided by saline wetlands:

Habitat Provision: Numerous plant and animal species, especially those that have adapted to high salt concentrations, rely on saline marsh ecosystems for survival. They safeguard habitats that are important for many species of animals, both marine and terrestrial. Wetland areas contribute to biodiversity in a number of ways, the most important of which are the supply of diverse habitats and the promotion of complex biological interactions.

Nursery Grounds: Saline wetlands are used as breeding grounds by many different types of aquatic species, including fish, crabs, and molluscs. These wetland regions provide shelter and an abundance of nutrients, making them perfect places for an organism to flourish and mature, and to gather the materials it needs to survive. The long-term viability of fish stocks and other aquatic population types depends on the availability of wetland regions suitable for use as nurseries.

Genetic Reservoir: Unique and highly specialised creatures that can survive in high-salt environments are often found in saline wetlands. These wetlands serve as genetic reservoirs, storing genetic material that can be used in future adaptations

Table 2.4 Examples of saline wetlands for supporting services

S.no.	Name of saline wetland	Country of location	Type of saline wetland	Supporting services ecosystem services provided
1	Chott el Jerid	Tunisia	Salt Flat	Nutrient cycling, sediment trapping, habitat for wildlife
2	Sambhar Lake	India	Salt Lake	Biodiversity habitat, sediment trapping, nutrient cycling
3	Great Salt Lake	USA	Terminal Lake	Wildlife habitat, sediment trapping, nutrient cycling
4	Lake Urmia	Iran	Endorheic Lake	Habitat for migratory birds, sediment trapping, nutrient cycling
5	Maranjab Salt Desert	Iran	Playa	Biodiversity habitat, nutrient cycling, sediment trapping
6	Etosha Pan	Namibia	Salt Pan	Wildlife habitat, nutrient cycling, sediment trapping
7	Lake Natron	Tanzania	Alkaline Lake	Flamingo breeding ground, sediment trapping, nutrient cycling
8	Salar de Uyuni	Bolivia	Salt Flat	Unique landscapes, wildlife habitat, sediment trapping
9	Lake Eyre	Australia	Salt Pan	Waterbird habitat, sediment trapping, nutrient cycling
10	Lop Nur	China	Salt Pan	Wildlife habitat, nutrient cycling, sediment trapping
11	Badwater Basin	USA	Salt Pan	Unique geological features, sediment trapping, nutrient cycling
12	Lake Abert	USA	Endorheic Lake	Bird habitat, sediment trapping, nutrient cycling
13	Lake Tuz	Turkey	Salt Lake	Bird habitat, sediment trapping, nutrient cycling
14	Koyashskoye Salt Lake	Russia	Salt Lake	Wildlife habitat, nutrient cycling, sediment trapping
15	Sebkha Sejoumi	Tunisia	Salt Pan	Bird habitat, sediment trapping, nutrient cycling
16	Lake Tengiz	Kazakhstan	Salt Lake	Waterbird habitat, sediment trapping, nutrient cycling
17	Lake Qarun	Egypt	Salt Lake	Fish habitat, sediment trapping, nutrient cycling
18	Kara-Bogaz-Gol	Turkmenistan	Lagoon	Bird habitat, sediment trapping, nutrient cycling
19	Soda Lakes of Kenya	Kenya	Saline Lakes	Waterbird habitat, sediment trapping, nutrient cycling
20	Lake Assal	Djibouti	Crater Lake	Unique geological features, nutrient cycling, sediment trapping
21	Oroumieh Lake	Iran	Endorheic Lake	Biodiversity habitat, sediment trapping, nutrient cycling

Table 2.4 (continued)

S.no.	Name of saline wetland	Country of location	Type of saline wetland	Supporting services ecosystem services provided
22	Salinas Grandes	Argentina	Salt Flat	Bird habitat, sediment trapping, nutrient cycling
23	Lake Uyuni	Bolivia	Salt Flat	Unique landscapes, wildlife habitat, sediment trapping
24	Lake MacLeod	Australia	Salt Lake	Waterbird habitat, nutrient cycling, sediment trapping

and so helping to sustain genetic variety. The genetic variety of a saline wetland ecosystem improves the ability of its species to adapt to changing environmental conditions and ensures their long-term survival.

Primary Production: In saline wetland environments, productivity can be kept at a primitive level mainly through the evolution of salt-resistant plant species. Plants found in wetlands, like the halophytes, have evolved to tolerate high salt. These plants are crucial for nutrient cycling and carbon sequestration. They transform solar energy into organic matter through photosynthesis, setting the framework for the food chain and supplying nourishment for the entire wetland ecosystem.

Soil Formation and Stabilisation: Saline wetland regions aid in the formation and consolidation of soil. Plants flourish in wetland areas thanks to the development of soil and the accumulation of organic matter, which in turn nourishes the soil with nutrients and organic material. The plants that thrive in wetland environments have deep roots that bind the soil, protecting it from erosion and keeping it stable. Long-term survival and efficient functioning of wetland ecosystems depend on these mechanisms.

Water Filtration and Retention: A natural filter, saline marsh areas remove sediments, fertilisers, and pollutants from water. The plants and soils in wetlands act as a filter, removing harmful substances from the water and preventing them from reaching downstream bodies of water. Wetlands not only help to retain and store water, which prevents excessive runoff and encourages groundwater recharge, but they also provide several other benefits.

Nutrient Cycling and Detoxification: The presence of salt marshes greatly facilitates nitrogen cycle and detoxification. This is accomplished through the uptake of excess nitrogen and phosphorus, which are subsequently transformed through microbial interactions with the plants. Wetland ecosystems regulate nutrient levels, decreasing the possibility of nutrient imbalances and the impacts of eutrophication in nearby water bodies.

Carbon Sequestration: The potential of saline wetland areas to collect and store carbon dioxide from the atmosphere makes them useful for carbon sequestration. Wetlands play an important role in mitigating climate change by absorbing carbon dioxide from the air and releasing it as oxygen. Long-term storage of carbon in wetland ecosystems helps with global climate management and keeping a carbon balance.

2.4.4 Cultural Ecosystem Services of Saline Wetlands

Saline wetlands provide various cultural ecosystem services that have significant value and importance for human societies. These services are related to the cultural, aesthetic, and spiritual connections that people have with wetland ecosystems which are mentioned in Table 2.5. Here are some examples of cultural ecosystem services provided by saline wetlands:

Recreation and Tourism: Birdwatching, nature photography, hiking, canoeing, and fishing are just some of the many popular outdoor activities that visitors can partake in at saline wetlands, making them a popular tourist attraction. Wetland ecosystems are a popular tourist destination for visitors who want to see beautiful landscapes, learn about local wildlife, and enjoy a wide range of outdoor pursuits. Economic benefits can accrue to local communities when visitors spend money at hotels, hire tour guides, and use other services directly related to tourism in wetland areas.

Cultural Heritage and Identity: Salt marshes often hold significant cultural and historical significance for the nearby towns. There's a chance they have roots in rituals practised in different cultures. For native people and other local populations, wetland areas have the ability to play an essential role in the preservation of cultural past and the provision of a sense of identity. There is a wide range of cultural practices and values thanks to the transmission of knowledge and understanding about wetland ecosystems from one generation to the next.

Artistic Inspiration: The unique topography, flora, and the way light plays across the surfaces of saline marshes all serve as inspiration for painters, writers, photographers, and other artists. The tranquillity and beauty of wetland environments make them ideal for generating ideas for works of art and other forms of creative expression. Wetlands are important to cultural and aesthetic expression because they have inspired numerous works of art, literature, music, and poetry.

Education and Research: Students and researchers can take advantage of saline wetlands as "living classrooms" because of their distinctive ecosystems. Having access to data on wetland ecology, biodiversity, and water quality makes these areas ideal for environmental education activities. Wetland regions are like nature labs in that they allow scientists to conduct research and experiments to better understand ecological processes and add to the body of scientific knowledge.

Spiritual and Sacred Sites: Salt marshes provide significant religious and spiritual significance for many religions and cultures. Many people associate these places with religious ceremonies, prayer, and the preservation of cultural traditions. Wetland regions have the potential to improve the mental and emotional health of individuals and entire communities by instilling a sense of awe, tranquillity, and oneness with nature.

Ecotourism and Nature-Based Therapies: Saline wetlands offer opportunities for ecotourism and nature-based therapies, such as eco-wellness retreats, meditation, and nature therapy programs. The peaceful and natural environment of wetlands can have a calming and rejuvenating effect on individuals, promoting

Table 2.5 Examples of saline wetlands for cultural services

S.no.	Name of saline wetland	Country of location	Type of saline wetland	Cultural services ecosystem services provided
1	Chott el Jerid	Tunisia	Salt flat	Cultural heritage, traditional salt harvesting, tourism
2	Sambhar Lake	India	Salt Lake	Cultural significance, traditional practices, tourism
3	Great Salt Lake	USA	Terminal Lake	Recreational activities, cultural significance, tourism
4	Lake Urmia	Iran	Endorheic Lake	Cultural heritage, traditional fishing, tourism
5	Maranjab Salt Desert	Iran	Playa	Cultural traditions, desert tourism, local practices
6	Etosha Pan	Namibia	Salt Pan	Ecotourism, cultural significance, traditional knowledge
7	Lake Natron	Tanzania	Alkaline Lake	Flamingo rituals, cultural practices, tourism
8	Salar de Uyuni	Bolivia	Salt Flat	Salt flats photography, cultural tourism, unique landscapes
9	Lake Eyre	Australia	Salt Pan	Aboriginal heritage, traditional practices, tourism
10	Lop Nur	China	Salt Pan	Historical significance, local culture, desert tourism
11	Badwater Basin	USA	Salt Pan	Geological wonders, cultural importance, tourism
12	Lake Abert	USA	Endorheic Lake	Bird watching, cultural heritage, photography tourism
13	Lake Tuz	Turkey	Salt Lake	Salt-related traditions, cultural practices, tourism
14	Koyashskoye Salt Lake	Russia	Salt Lake	Local customs, cultural significance, tourism
15	Sebkha Sejoumi	Tunisia	Salt Pan	Cultural heritage, traditional salt extraction, tourism
16	Lake Tengiz	Kazakhstan	Salt Lake	Local traditions, cultural significance, birdwatching
17	Lake Qarun	Egypt	Salt Lake	Traditional fishing, cultural practices, birdwatching
18	Kara-Bogaz-Gol	Turkmenistan	Lagoon	Cultural heritage, local practices, birdwatching
19	Soda Lakes of Kenya	Kenya	Saline Lakes	Local rituals, traditional practices, birdwatching
20	Lake Assal	Djibouti	Crater Lake	Geothermal features, cultural practices, desert tourism
21	Oroumieh Lake	Iran	Endorheic Lake	Cultural traditions, traditional fishing, tourism
22	Salinas Grandes	Argentina	Salt Flat	Cultural heritage, traditional customs, photography tourism
23	Lake Uyuni	Bolivia	Salt Flat	Salt flat photography, cultural tourism, unique landscapes
24	Lake MacLeod	Australia	Salt Lake	Indigenous heritage, traditional practices, tourism

mental and physical well-being. Wetlands provide spaces for relaxation, reflection, and connecting with nature, supporting holistic health and wellness.

Cultural and Culinary Traditions: Especially with regard to traditional food and medicinal plants, saline wetlands enrich cultural and culinary traditions. Wetland environments are a great place for traditional fishing methods, the collection of food plants, and the harvesting of salt. Traditional healing and cooking methods that make use of wetland ingredients help to maintain and perpetuate local cultures.

2.5 Threats to Ecosystem Services of Saline Wetlands

Threats to the ecosystem services provided by saline wetlands are numerous, just as they are in other types of ecosystems. The ecological benefits of salt marshes are threatened by a number of factors.

Habitat Loss and Fragmentation: One of the greatest threats to saline wetland ecosystems today is the destruction and fragmentation of habitats brought about by human activities including urban development, agriculture, the construction of infrastructure, and industrial operations. Wetlands are an important home for many species, and if they were converted into other sorts of land uses, it would destroy this habitat and disrupt natural processes (Zheng 1997).

Pollution and Contamination: When it comes to pollution and contamination, salty wetland ecosystems are particularly vulnerable. Agricultural runoff, industrial discharges, and improper waste management are just a few of the many potential entry points for pollutants and toxins into wetland ecosystems. These pollutants may degrade water quality, threaten the survival of aquatic animals, and disrupt the natural processes that keep salt marshes healthy ecosystems.

Invasive Species: Invasions of non-native plant and animal species pose a significant threat to saline wetland ecosystems. It is possible for invasive species to affect the structure and function of wetland ecosystems by outcompeting native species and thereby disrupting the ecological equilibrium. They may reduce biodiversity, reduce the quality of the habitat, and otherwise affect the ecological benefits that salt marshes provide.

Overexploitation and Unsustainable Use: Unsustainable harvesting of natural resources has the potential to deplete natural resources and disrupt the delicate balance of saline wetland ecosystems. Some common instances are unsustainable tourism practices, excessive salt extraction, and overfishing. Overexploiting saline wetlands can lead to the extinction of key species and the disruption of biological interactions, reducing the wetlands' potential to provide essential ecosystem services.

Climate Change: The effects of climate change, such as rising sea levels, warmer temperatures, different precipitation patterns, and more intense storms, are especially harmful to saline marshes. Climate change has the potential to affect the distribution of animals and plants by, among other things, causing saltwater intrusion, habitat loss, changes to hydrological regimes, and species extinction.

These alterations may affect the ability of saline wetland ecosystems to provide ecosystem services while also threatening their integrity and functionality.

Infrastructure Development: The disruption of salt marshes' natural hydrological regimes can occur when man-made structures like dams, dykes, and roads are built in the area. Water supply, nitrogen cycling, and wetland functionality are all susceptible to alterations in water flow patterns. Habitat fragmentation, caused by increased human development, hinders the movement of animals and alters the equilibrium of ecosystems.

Lack of Awareness and Conservation Efforts: The success of conservation efforts may be hampered by people's inability to recognise and comprehend salt marshes' worth and significance. If proper conservation measures are not taken and protective legislation is not enforced, saline wetland habitats are especially vulnerable to decline and the loss of ecosystem services.

2.6 Threats to Saline Wetland Ecosystems

Saline wetland ecosystems face numerous threats that can impact their integrity, functionality, and biodiversity which are diagrammatically represented in Fig. 2.5.

2.6.1 Habitat Loss and Degradation

Agriculture, urban growth, infrastructure projects, and industrial operations all contribute to the loss of wetland habitats and their fragmentation. This leads to the destruction of wetland ecosystems. Wetland ecosystems are also threatened by drainage and land reclamation projects. The ecological integrity and biodiversity of

Fig. 2.5 Threats to saline wetlands

saline wetland ecosystems are in jeopardy due to habitat loss and degradation. Here some of the most crucial factors that have led to the destruction of saltwater wetlands are given:

Land Conversion: Salt marshes have been altered for agricultural, urban, and industrial development, which has contributed to their demise. Human actions, such as draining and filling wetlands to make place for other purposes, frequently lead to the degradation of wetland ecosystems.

Drainage and Alteration of Hydrological Regimes: Because of their unique ecological characteristics, wetlands can only survive in the presence of water. Adjustments to natural hydrological regimes, such as damming and channelisation, have the effect of interrupting the flow of water in saline wetlands, as does the draining that is done for agricultural purposes. The disruption of natural systems and the resulting loss of biodiversity can result from this.

Fragmentation: Fragmentation of salt marshes occurs when human activity, such as the installation of infrastructure or the conversion of land, divides wetland habitats into smaller, more isolated portions. Fragmentation reduces the size and connectivity of wetland ecosystems, limiting the range of species and causing havoc with natural processes.

Pollution: Pollution from a wide variety of sources, including factories, farms, and cities, poses a significant threat to salt marshes and other saline wetland ecosystems. Wetland water quality can be compromised by chemical contaminants and excess nutrients like nitrogen and phosphorus, which threaten the wetland's aquatic life and plants.

Invasive Species: The ecological balance that has been maintained in salt marshes for millennia could be disrupted by the introduction of non-native plant and animal species. Invasive species are a leading cause of wetland habitat degradation because they alter the composition of vegetation and induce structural disruptions in wetland ecosystems by outcompeting native species for resources.

Climate Change Loss of salt marsh habitat and its quality are suffering serious consequences as a result of global warming. When temperatures continue to rise, altered precipitation patterns occur, and sea levels continue to rise, it has an effect on the hydrological cycles and salinity levels in wetlands. Vegetation loss, habitat reduction or loss, and heightened vulnerability to other stresses are all possible outcomes of these alterations.

Soil Erosion: Soil erosion, carried on by farming methods, clearing of forests, or poor land management, can cause sedimentation inside salt marshes. Sedimentation can have negative effects on a wetland's environment by lowering water quality, smothering aquatic plants and animals, and upsetting the delicate balance that exists there.

Overgrazing: Excessive grazing by cattle in and around salt marshes can lead to degradation of wetland habitats. Overgrazing by cattle reduces wetland biodiversity and impairs the ecosystem's ability to provide essential services by compacting the soil, altering the flow of water, and trampling vegetation.

2.6.2 Water Extraction and Alteration

Redirecting the water supply for activities such as irrigation, domestic consumption, or industrial production can disrupt the normal hydrological cycles of salt marshes. Reducing water availability, altering salinity levels, and upsetting wetland ecosystems are among side effects of water abstraction. Water diversion and manipulation have the ability to severely disrupt the hydrological cycles, salinity levels, and biological functioning in salty wetland habitats. Below, we explore the implications of water extraction and change as a threat to salt marshes.

Disruption of Natural Hydrological Regimes: Unique ecological characteristics of saline wetland habitats are maintained by a variety of water regimes, including seasonal fluctuations in water level and the influence of the tides. However, these natural hydrological regimes can be disrupted if an excessive amount of water is withdrawn for agricultural purposes, industrial uses, or urban expansion. Depending on the severity of the disturbance, the wetland ecosystem may experience a decline in water supply, a shift in water flow patterns, or a degradation in water quality.

Reduction in Water Levels: When water is taken out of a salt marsh, the water table drops, which in turn lowers the total amount of water in the system. The presence or absence of water has a direct effect on the amount of suitable habitat for aquatic organisms, especially those that have evolved to live at a particular water depth. It may also alter the viability of wetland vegetation, leading to changes in plant communities and a reduction in the variety of species present.

Altered Salinity Levels: The high salinity levels found in salty wetland environments are a telltale sign of their uniqueness. Plant and animal life that can be found there are consequently quite specialised and well adapted to the conditions there. However, water withdrawal has the potential to upset the delicate balance of salinity in the marsh. Some plant and animal species may not be able to survive if the salinity of the remaining water rises due to the diversion of freshwater supplies. However, reducing the salinity of a body of water by draining out salty water can affect the survival of salt-tolerant creatures.

Impaired Nutrient Cycling: Nutrient cycling relies in large part on the hydrological cycles that occur in saline wetlands. This mechanism facilitates the transfer and redistribution of nutrients critical to the well-being of wetland vegetation and aquatic organisms. Reduced availability of nutrients and a shift in the ecosystem's nutrient balance are both consequences of water withdrawal disruptions in these cycles. This disturbance may have an adverse effect on the wetland's ability to produce and function, which in turn may have an impact on the growth and survival of its plant and animal inhabitants.

Ecological Consequences: Changes in salinity and hydrological regimes in saline wetlands can have far-reaching consequences for the entire ecosystem. Plant communities can fluctuate as a result of changes in water availability and water salinity, favouring some species over others. Changes in vegetation can have an influence on herbivores, which can have an effect on higher trophic levels; hence, these alterations may alter the dynamics of the food web. The loss of plant and animal species that have adapted to life in salty settings has the potential to disrupt ecological webs and reduce biodiversity overall.

Impact on Migratory Species: Important nesting and resting locations for migrating birds and other wildlife can be found in saline marshlands. When water is diverted elsewhere, it can alter water levels and salinity, which in turn can affect the accessibility of suitable habitats and disrupt the life cycles of these species. Less water means fewer food sources and fewer locations to nest, which can have a severe effect on breeding and migration success.

2.6.3 Pollution

Agricultural runoff, industrial discharges, and untreated sewage are only a few of the many potential sources of contamination in saline wetland habitats. Excessive concentrations of nutrients, heavy metals, and pesticides are only some of the contaminants that can result from these pollution sources. Pollution has a negative effect on aquatic life, plant life, and water quality. Pollution poses a serious threat to salt marshes because it degrades water quality, harms ecosystem health, and reduces biodiversity. Saline wetlands are particularly vulnerable to the negative effects of pollution due to their unique ecological characteristics. The following is a look at how pollution has harmed salt marshes and what may be done about it.

Agricultural Runoff: Pollution of salt marshes and other saline ecosystems is exacerbated by agricultural runoff containing fertilisers, pesticides, and herbicides. Excessive use of these pesticides on the farmland around the wetland might cause nutrient enrichment, especially of nitrogen and phosphorus, in the wetland's ecosystem. The wetland's plant and animal communities could be negatively impacted by the nitrogen flow because it could lead to algae blooms, lower oxygen levels, and upset the wetland's ecological balance.

Industrial Discharges: Industries located in close proximity to salt marshes may emit pollutants such as hazardous chemicals, heavy metals, and organic compounds into bodies of water. Wetland ecosystems are vulnerable to these poisons because they can enter the water supply through discharge or runoff, reducing water quality and raising the risk to aquatic life. Contaminants accumulating in sediments and organisms can have both immediate and long-term effects on the surrounding ecosystem.

Urban and Domestic Waste: One cause of contamination in salt marshes is the careless disposal of municipal and domestic trash. Untreated sewage, garbage, and other forms of waste that end up in water bodies pose a health danger to the species that call wetlands home and can have a severe effect on the quality of the water itself. Human waste introduces new diseases and creates environments favourable to the growth of harmful germs, which can have an impact on the wetland's species and the humans who utilise it.

Oil and Chemical Spills: There may be devastating consequences for the ecology if oil or dangerous chemicals are released into salt marshes, whether by mistake or on purpose. Accidents on the road, manufacturing errors, and illegal disposal are just some of the many potential causes of such disasters. The natural cycle of the ecosystem is disrupted when oil and chemical spills form a film that covers the vegetation of wetlands, putting the health of aquatic species at risk. Long-term, the effects might be catastrophic, leading to the destruction of habitat and the spread of toxins.

Sedimentation: Sedimentation is a form of pollution that can damage saline wetland ecosystems; it is typically caused by soil erosion from neighbouring places. When silt accumulates in water bodies, it can smother benthic species, block light from reaching the water, and reduce the wetland's ability to exchange oxygen. This might potentially affect the structure of the wetland environment, hinder the migration of aquatic organisms, and reduce the growth of aquatic vegetation.

Plastic and Debris: The contamination of saline wetland ecosystems by plastics and other trash is a serious problem. Microplastics, which are too small to be considered full particles of plastic, have the potential to accumulate in the ecosystem of wetlands because plastic does not biodegrade. Marine creatures and birds may confuse plastic trash for food and ingest it or become entangled in it. Garbage not only detracts from the wetland's aesthetic value but also poses a threat to the health of the surrounding environment.

2.6.4 Invasive Species

Saline wetland ecosystems are delicately balanced, thus it is important to prevent the introduction of non-native plant and animal species. An ecosystem's natural processes can be thrown off by the introduction of an invasive species because it can outcompete native species for food and other resources. Invasive species pose a serious challenge to saline lakes, threatening their ability to maintain ecological stability, native biodiversity, and ecosystem health. Saline lakes are particularly vulnerable to invasion because of the unique circumstances that prevail in these bodies of water.

Competitive Exclusion: It's not uncommon for invading species to outcompete native ones for food, habitat, and even physical space. Competitors to native species are generally invasive species. Under saline lakes, they can quickly establish themselves and take over ecological niches, displacing native species that have evolved to thrive in the lake's unique salinity conditions. They are able to do this due to their greater reproductive capacities, which allow them to outcompete native species. Extinction of native species, loss of biodiversity, and disruption of ecological processes are all possible outcomes of competitive exclusion.

Altered Food Webs: The natural food webs that exist inside salt lakes could be disrupted by invasive species. Population dynamics and trophic interactions may change as a result of their predation or competition with native species for food sources. There may be domino consequences across the ecosystem if food webs are altered, affecting the populations and distributions of other species and possibly causing ecological instability.

Habitat Modification: The introduction of invasive species into saline lake habitats has the potential to alter the physical structure of those ecosystems. Some invasive aquatic plants, for instance, can alter the make-up and structure of submerged vegetation by forming dense mats or covering large areas. The loss of suitable habitat for native species, changes to the lake's water circulation patterns, and a disruption of the ecosystem's delicate biological balance are all possible outcomes of this habitat modification.

Hybridisation and Genetic Pollution: When invasive species mate with native ones, it can dilute the gene pool of the native population. This has the potential to cause the extinction of local species. Hybridisation can cause local adaptations to be lost, fitness to decline, and the emergence of hybrid swarms that can then outcompete and extirpate native populations. Long-term repercussions of this genetic pollution on native species' capacity to maintain population sizes and adjust to changing climatic and environmental conditions are possible.

Altered Nutrient Cycling: Nutrient cycle mechanisms in saline lakes may shift if invasive species are present. The nutrition cycle could be thrown off if introduced species have differing nutrient needs or absorption capacities compared to native ones. Changes in nutrient availability, eutrophication, and water quality are just some of the ways this might affect the biological functioning of a saline lake.

Introduction of Pathogens and Parasites: Threatening native species, and invading species can operate as reservoirs for pathogens and parasites in salty lakes. Threatening the health and survival of native species, these exotic viruses and parasites have the capacity to infect native organisms. Furthermore, invading species pose a threat to native species because they may introduce parasites or illnesses to an ecosystem for which native species have no defences.

2.6.5 Climate Change

Rising sea levels, altered precipitation patterns, and rising temperatures are the greatest threats to saline wetland ecosystems. Changes in salinity, hydrological cycles, species distribution, and ecosystem health are all potentially affected by these alterations. Climate change poses a serious threat to salt marsh ecosystems due to its impact on hydrology, biodiversity, and ecological functioning. The unique characteristics and increased sensitivity of saline wetlands make them particularly vulnerable to the adverse effects of climate change.

Rising Temperatures: Salt marshes may be significantly impacted by the increasing average global temperature that is a result of climate change. Rising temperatures can increase the rate of evaporation from wetlands, reducing the amount of water available within the wetland. Wetland plants and aquatic animals, especially those adapted to specific salinity levels and water depths, may be negatively impacted by this drop in water level.

Changes in Precipitation Patterns: Precipitation patterns are shifting as a result of climate change, which is having an impact on the frequency, intensity, and location of downpours. The hydrological cycles and equilibrium between the amount of freshwater that enters saline wetlands and how salty the conditions are are based on certain rainfall patterns. The water supply, salt levels, and general productivity of wetland ecosystems are all negatively impacted by changes in precipitation cycles.

Sea-Level Rise: Sea-level rise due to climate change is a major threat to coastal wetlands and salt marshes. Saline wetlands along the coast are at risk of being flooded and having saltwater seep into them due to rising sea levels. An increase in salinity, the loss of habitat, and the spread of salt-tolerant plant and animal species are all possible outcomes of saltwater's incursion inland. As these wetland varieties provide vital ecological services like coast protection and carbon storage, their loss can have far-reaching repercussions on the environment and economy.

Altered Storm Patterns: Climate change can alter storm patterns, leading to more intense and frequent storms in some regions. Storms of this magnitude have the ability to alter coastal environments by eroding and flooding coastal areas. Damage to the salt marsh's physical framework and integrity can result in the loss of habitat, disruptions in hydrological connectivity, and the introduction of pollutants and sediment.

Shifts in Species Distribution: The distribution of plant and animal species in salt marshes may shift as a result of climatic changes. As environmental conditions worsen in their native habitat, some species may be compelled to move to other areas with more favourable temperatures or salinity. These changes may cause irreparable harm to ecosystems by upsetting the balance of natural communities and possibly even leading to the extinction of some species that have evolved to thrive in their current ecosystem. The distribution of plant and animal species in salt marshes may shift as a result of climatic changes. As environmental circumstances worsen in their native range, some species may be compelled to seek out new homes in areas with more favourable salinity or temperature ranges. These changes may lead to the extinction of species that have evolved to thrive in their current ecosystem, as well as to the disruption of other important ecological connections.

Loss of Carbon Storage: Salty and brackish wetlands play a crucial role in carbon sequestration and storage. They contribute to climate management because of their capacity to absorb and store large amounts of carbon dioxide. However, saline wetland ecosystems are vulnerable to degradation and loss due to climate change consequences including reduced water supply and increased salt. In addition to contributing to the quickening of climate change, this loss releases carbon that had been trapped below.

2.6.6 Overgrazing

Allowing cattle to graze in and among salt marshes can cause soil compaction, plant trampling, and a disruption in the normal flow of water. The destruction of habitat and the loss of biodiversity are both possible results of overgrazing. Overgrazing saline lakes and their surrounding ecosystems is a serious threat that may be traced back to a wide range of negative ecological and environmental outcomes. In the following paragraphs, we will examine the effects of overgrazing as a threat to salt lakes.

Vegetation Degradation: Overgrazing by cattle may be to blame for the devastation of the vegetation around salt lakes. When animals are given free rein to graze, they often destroy vegetation at a faster pace than it can regenerate. Vegetation is stripped away, soil is eroded, and native plant species lose critical habitat as a result of this process. Reducing the amount of vegetation in an area can make it more open to colonisation by invasive species, which can further disrupt the ecological balance of the area.

Soil Erosion: In areas where soils are already fragile, such as those around salt lakes, excessive grazing can exacerbate the problem. Animals constantly grazing and trampling the soil makes it more porous, making it more vulnerable to wind and water erosion. Sediment introduced into the lake as a result of erosion can have negative effects on water quality and interfere with the ecosystem's natural processes.

Nutrient Imbalances: Overgrazing can disrupt the natural processes that maintain nutrient cycling in the area around salt lakes. Overgrazing can enhance soil with nutrients because of the localised accumulation of animal excrement. As a result of this enrichment, nutritional imbalances may develop, altering the composition of plant communities and possibly facilitating the spread of invasive species. Nutrient imbalances can have a severe effect on the water quality of the salty lake by promoting eutrophication and the establishment of hazardous algal blooms.

Loss of Biodiversity: Overgrazing of the vegetation around the salt lake can have a negative impact on the ecosystem's biodiversity. Heavy grazing pressure threatens the survival of local plant species that have evolved to thrive in the area's specific climate and soil. Vegetation loss can have an adverse effect on the surrounding fauna, such as insects, birds, and other animals that rely on particular plant species for sustenance and refuge. The loss of biodiversity has the potential to disrupt ecological systems and undermine the planet as a whole.

Water Quality Impacts Water quality in salt lakes can deteriorate for a variety of reasons, and excessive grazing is one of them. When vegetation is cleared away, the ground around the water loses some of its natural filtration capacity, allowing more sediment, nutrients, and contaminants to make their way into the water. A decline in clarity, an increase in turbidity, and a shift in the chemical makeup of the water are all possible results of an invasion of contaminants. The bad water quality can have a devastating effect on aquatic creatures, threatening their health and the ecosystem's ability to function normally.

Disruption of Hydrological Cycles: Too much overgrazing could disrupt hydrological cycles in the region around salt lakes. Reducing evapotranspiration rates due to a lack of vegetation leads to shifts in local rainfall patterns and the availability of water. Water levels, salinity, and the availability of suitable habitats for aquatic creatures are only some of the ways in which this disruption of hydrological cycles could alter the salt lake's ecosystem.

2.6.7 Saltwater Intrusion

Human activities such as groundwater depletion or altered coastal hydrology can cause saltwater to enter inland saline wetlands. Plant life and ecological processes in these areas might be negatively impacted by a rise in salinity. Coastal saline wetland ecosystems are especially susceptible to the negative effects of saltwater intrusion. When discussing wetland ecosystems, the phrase "saltwater intrusion" is used to describe the penetration of salty water into normally freshwater sources, such as aquifers or surface water. The water's salinity rises as a result of this phenomenon.

Changes in Hydrology: When saltwater is introduced to saline wetlands, it disrupts the ecosystem's delicate hydrological balance. When saline water seeps into freshwater aquifers or enters salty water bodies on the surface, the salinity gradient inside the wetland shifts. This shift in salinity has the potential to affect water flow patterns, which could result in changed circulation, reduced freshwater inflows, and different water levels. The distribution of wetland plants is one aspect that can be affected by changes in hydrology, as can the reproductive cycles of aquatic animals and the ecological processes that take place within the wetland ecosystem as a whole.

Loss of Freshwater Resources: Freshwater supplies can be depleted when saltwater is present in saline wetlands. The availability of freshwater to the plants and animals of wetlands is reduced when saltwater is introduced into freshwater aquifers. There may be changes to the wetland's species composition if freshwater supplies are reduced, as this can lead to stress and mortality in freshwater-dependent species.

Impacts on Vegetation: Saltwater intrusion can be detrimental to the growth and survival of vegetation that has adapted to the specific salinity levels seen in saline wetland environments. The ability of wetland plants to absorb water and nutrients from their surroundings is hampered when saline water seeps into their root zones. This physiological stress can slow plant growth, make plants more susceptible to pests and diseases, and eventually cause marsh vegetation to perish. Vegetation loss can have far-reaching consequences for other forms of life, such as those that depend on wetland plants for sustenance and shelter.

Changes in Species Composition: The introduction of saltwater can have a dramatic effect on the species composition of saline wetland ecosystems. Increases in salinity favour the growth of salt-tolerant species while threatening the extinction of others that must rely on freshwater. The loss of biodiversity, the simplification of ecological systems, and the alteration of their interdependencies are all possible outcomes of this shift in species composition. It could also promote the spread of exotic species that thrive in saltier environments. The native species and the wetland's ecological balance would suffer further as a result of this.

Disruption of Ecosystem Services: When saltwater enters into saline wetlands, it threatens to undermine the ecosystem services offered by these areas. Protecting coastal areas, filtering water, and providing a home for a wide variety of animals are just a few of the many vital functions performed by wetland ecosystems. If the salinity levels in the wetland continue to rise, the wetland's ability to prevent coastal erosion, filter pollutants, and provide a habitat for a diverse array of animal species could be compromised. Reduced coastal resilience, diminished water quality, and fewer opportunities for recreation are only some of the societal and economic costs associated with the loss or degradation of key ecosystem services.

2.6.8 Agricultural Practices

Nutrient runoff from nearby agricultural areas that make excessive use of fertilisers, pesticides, and irrigation can contaminate salt marshes. Algae blooms, oxygen deprivation, and ecological disruption are only some of the potential outcomes of nutrient enrichment. When farming practices include improper land management and the overuse of agricultural pesticides, they pose a significant threat to salt marshes. Below, we explore the ways in which agricultural practices pose a threat to salt marshes.

Nutrient Runoff: It is usual practice in agriculture to employ fertilisers, which promote rapid plant growth. An excessive amount of fertiliser on a field, however, might cause contamination of nearby water sources, especially salt marshes. The excessive growth of algae and aquatic plants known as eutrophication, which can be caused by this nitrogen flow, depletes the water of oxygen. To put it simply, eutrophication is bad for wetland ecosystems because it reduces biodiversity, wipes off fish populations, and lowers water quality.

Pesticide Contamination: Farmers regularly use agrochemicals like insecticides and herbicides to prevent damage to their crops from pests and weeds. However, improper application or excessive usage of these chemicals can cause contamination of the local ecosystem through surface runoff or leaching into groundwater. Pesticide runoff in water supplies is a threat to aquatic life, particularly in salt marshes, and has the potential to reduce biodiversity and alter ecological processes.

Soil Erosion: Agricultural practices that are harmful to the environment, such as intensive tilling and the destruction of natural vegetation for crop production, can lead to soil erosion. If the soil in agricultural areas is allowed to deteriorate, the topsoil could end up in nearby bodies of water, especially salt marshes. Sedimentation, an increase in turbidity, and even changes in the quality of the water within the wetland can all arise from erosion of the soil. Sedimentation can alter the wetland's physical structure, which can suffocate vegetation and affect the habitat of aquatic organisms.

Water Extraction: In the course of specific farming practices, it may be necessary to pump water from the ground or other nearby sources for use in irrigation. Large-scale abstraction threatens water supplies, especially the freshwater that carries salt out of wetland areas. Reduced water availability can lead to changes in salt levels, altered hydrological patterns, and the loss of wetland plants and species, all of which are in turn dependent on adequate water sources.

Wetland Drainage and Conversion: Due to the excessive salinity, salt marshes are sometimes drained or converted into agricultural land. Wetland drainage is the action of draining water from a wetland ecosystem. Creating ditches and other drainage channels is a common method for achieving this goal. This method alters the wetland's hydrology, leading to salinity changes and the extinction of species that rely on wetland ecosystems for survival. Wetland conversion is the process by which wetlands are transformed into farmland. Wetland ecosystems are destroyed in this way, and the vital ecosystem services they provide are also lost in the process.

Genetic Pollution: Genetic contamination from agricultural operations, such as the manufacturing of genetically modified crops, poses a serious threat to saline wetland ecosystems. Pollen or seeds from genetically modified crops can contaminate local wild plant populations, particularly those found in salt marshes, if they are blown or washed away by the wind or water. Genetic pollution can threaten the delicate ecological balance of wetland ecosystems by causing the loss of genetic diversity and the replacement of local plant species.

2.6.9 Dams and Water Infrastructure

Construction of dams, levees, and other water control structures can alter natural water flows, alter sediment deposition patterns, and fragment salty wetland ecosystems' natural habitats. The ecology and biodiversity of wetlands could be negatively impacted by these alterations. When dams and other water infrastructure disrupt the natural flow of water and disrupt the ecosystem's delicate hydrological equilibrium, they represent a significant threat to salt marshes. The following discusses the hazards posed to salt marshes by dams and other types of water infrastructure:

Alteration of Water Flow: The development of dams and other types of water infrastructure can change the natural flow patterns of rivers, streams, and other water sources that feed into salt marshes. Reduced or altered freshwater inflows into the wetland as a result of water flow restriction might affect salt levels and the hydrological dynamics of the ecosystem. Alterations to the water's flow can disrupt the natural flooding and draining cycles, which in turn can affect the growth and reproduction of wetland plants and the accessibility of suitable habitats for aquatic species.

Sediment Trapping: There is less sedimentation further downstream because dams trap the sediments carried by the water. The deposition of sediment is crucial to the maintenance of saline wetlands, especially those located near river deltas or estuaries, as it helps to maintain the wetland's physical structure and allows for the formation of wetland vegetation. Dams can reduce sediment delivery, leading to issues including erosion, land subsidence, and the loss of wetland areas.

Fragmentation of Habitat: The normal flow of water can be interrupted by dams, isolating salt marshes from their upstream and downstream environments. Fish, amphibians, and invertebrates, all of which rely on the connectivity of water systems in order to spawn and access suitable habitats, can have their migration hampered by this fragmentation. Loss of genetic diversity, decreased population reproduction, and ecological imbalances are all possible outcomes of reducing the wetland ecosystem's connectivity.

Changes in Salinity: The salinity of salt marshes may shift as a result of construction of dams and other water structures. Depending on the management practices in place, dams can cause either an increase or decrease in salinity. This is because dams regulate both the quantity and quality of freshwater supplies. Saline wetlands may be negatively affected by drastic changes in salinity because their inhabitants have acclimated to specific salinity levels. This could change the composition of aquatic habitats and the range and density of wetland plants.

Loss of Floodplain Functionality: The large numbers of salt marshes that populate floodplains serve as de facto reservoirs for the water that flows across them at peak times. Dams can control and reduce flood occurrences, which can lead to the loss of natural floodplain functionality. When floods are reduced, the normal ecological processes that keep wetland areas productive, like sediment deposition and nutrient enrichment, might be disrupted.

Loss of Ecosystem Services: Saline wetlands are very valuable because of the ecosystem services they provide, including water purification, reduced flood danger, and habitat provision. Dams and other forms of water infrastructure can alter the natural flow of water, potentially disrupting these functions. If the amount of freshwater entering an ecosystem decreases, the salinity of the water may rise, and wetlands will no longer be able to filter harmful chemicals out of the water. The wetland's ability to operate as a natural buffer against extreme weather events is diminished, and the risk of downstream flooding is raised when floodplain functionality is lost. Loss of floodplain functionality can enhance both of these dangers.

2.6.10 Mining and Extraction

Extraction of minerals, salts, or other resources from saline wetlands can lead to habitat damage, soil compaction, and water pollution. Wetland habitats can be directly degraded and nearby areas can be disrupted by mining operations. Since mining and extraction often involve the removal and modification of land, vegetation, and water supplies, they represent significant threats to salt marshes. This increases the likelihood of harm to salt marshes. The following is a look at the potential risks posed to salt marshes by mining and extraction:

Habitat Destruction: Salt marsh and wetland ecosystems are particularly vulnerable to destruction due to mining and extraction activities. Wetlands may be drained, excavated, or have vegetation removed to expose mineral deposits or extract resources. Marsh vegetation and the breeding and feeding sites for a wide variety of animals and plants are lost as a result of the degradation of this ecosystem. The loss of wetland ecosystems can reduce biodiversity and cause ecological imbalances if it forces the migration or extinction of species that rely on these habitats.

Water Pollution: Heavy metals and solvents used in mining and extraction processes can seep into nearby water sources and pose a threat to salt marsh ecosystems. Water pollution and a decline in water quality can happen from runoff and seepage from mining sites carrying pollutants into wetland areas. Water contamination has negative effects on aquatic organisms' health and survival, can disrupt food chains, and can pose hazards to human health when utilised as a drinking water source or when wetlands are used to sustain local livelihoods.

Changes in Hydrology: Mining and extraction activities pose a risk to the hydrological processes that sustain saline wetland habitats. By excavating, dewatering, and constructing drainage systems, it is possible to change water flow patterns, redirect watercourses, and reduce water tables. These changes disrupt the equilibrium of the wetland's water levels and flows, leading to new hydrological regimes, less water availability, and even the potential for some wetland sections to dry out entirely. Wetland vegetation, animal habitats, and the ecosystem's ability to operate as a whole could all suffer as a result of these changes.

Sedimentation and Erosion: Erosion and sedimentation of salt marshes and other related habitats can be exacerbated by mining and other forms of extraction. The loss of vegetation and the disturbance of the soil can hasten the processes of erosion, which in turn can cause the deposition of silt in nearby water bodies. Sediment buildup can smother aquatic ecosystems, reduce the amount of light reaching the water's surface, and impede the reproductive success of aquatic animals. Erosion can worsen sedimentation and further impair the wetland environment if vegetation and soil stability are lost in wetland zones.

Disruption of Ecosystem Services: Important ecosystem services supplied by saline wetlands include water filtration, nitrogen cycling, and the provision of habitat. The ability of wetlands to sustain biodiversity and provide benefits to communities in their immediate area is at risk due to the possible disruption of these services caused by mining and other extraction operations. Degradation of the wetland's water purification capacity due to contaminated water can reduce its ability to filter out contaminants and maintain stable water quality. The wetland's capacity to provide habitat for plants and animals may be diminished as a result of hydrological changes and habitat loss. The ecosystem's ability to store carbon, control floods, and provide recreational and cultural benefits could all suffer as a result of these changes.

2.6.11 Overfishing

Overfishing and other destructive fishing practices threaten saline marsh fish populations to the point of extinction. The loss of key fish species in a wetland ecosystem can disrupt food webs and alter the complex biological relationships that sustain life there. Overfishing is a major threat to aquatic ecosystems that are interconnected with salt marshes. Saline wetland ecosystems serve as critical spawning, nursery, feeding, and migrating areas for a wide variety of fish species. It's not uncommon to find saline wetland ecosystems along the ocean or in an estuary. Overfishing is a threat to salt marshes, and its effects are detailed below.

Disruption of Food Chains: The delicate equilibrium of marine food systems within salt marshes could be disrupted by intensive fishing. Unsustainable rates of fish removal can alter the dynamic between predators and prey, which can have far-reaching consequences for the entire ecosystem. Changes in the content and structure of an ecosystem can result, for instance, from a reduction in the number of top predator species, which could lead to an increase in the number of prey populations.

Decline in Fish Populations: Overfishing may be to blame for the decline of salt marsh fish populations. In the absence of proper laws and management, fish stocks can be depleted through the practice of targeted fishing of commercially valuable species. Fish play important roles in nutrient cycling, energy transmission, and maintaining the health of the ecosystem as a whole, therefore a decline in their numbers could have a negative impact on these processes.

Altered Community Structure: Changes in the composition of salt marsh communities may come from excessive fishing. It is possible for the abundance and variety of other species in an ecosystem to shift after the removal of certain fish species. Changes in the prevalence of prey species, variations in the dynamics of competitive interactions, and even probable population declines in non-targeted species are all possible outcomes of such interventions.

Impacts on Ecosystem Services: Many ecological services, such as fishery production, nutrient cycling, and water purification, rely on saline wetland areas. Overfishing can diminish these benefits since it cuts down on the number of fish accessible for both commercial and subsistence fisheries. If fish populations continue to decline, communities that rely on fishing as a key source of revenue may face economic difficulties. In addition, the wetland's water quality and ecosystem health could be impacted by the decline in fish biomass, which could have repercussions for the wetland's nutrient recycling activities.

Disruption of Spawning and Reproduction: Many types of fish are limited to breeding in saltwater or brackish wetlands. Overfishing may have a negative effect on the reproductive success of these species because it reduces the number of mature individuals available for reproduction. The wetland's capacity to sustain fish populations over the long run may be jeopardised if this disturbance reduces the number of juveniles entering the population.

Impacts on Other Wildlife: Not only do fish require saline wetland habitats, but so do a wide range of other animals, such as numerous species of birds, reptiles, and mammals. Overfishing has the potential to negatively affect these species since it reduces the availability of prey. Reducing the amount of fish in a region may disrupt the balance of the wetland's biological interactions and linkages. This is because there will be fewer prey for these animals to eat.

2.6.12 Recreation and Tourism Pressure

The degradation of salt marsh habitat, including the trampling of plants, the pollution of rivers, and the disturbance of animal population dynamics, can be caused by large numbers of people and unmanaged recreational activities. Wetland habitats are especially vulnerable to human impact because of poor waste management by visitors. The local ecosystem may suffer if salt marshes are subjected to excessive pressure from recreational and tourist activities. While engaging in these activities might be good for the economy and environmental awareness, they can have a negative impact on the wetland ecology if not properly regulated and conducted using sustainable practices. The following is an examination of the dangers that recreational pursuits and visitors represent to salt marshes:

Habitat Degradation: Spending too much free time in salt marshes and their environs could have negative environmental consequences. Wetland vegetation loss, trampling of vulnerable habitats, and soil erosion are all potential outcomes of camping, hiking, and the use of off-road vehicles in poorly managed places. The destruction of delicate ecosystems can also lead to these issues. If the water quality continues to decline, it could have a negative impact on the local aquatic ecosystem. The inappropriate disposal of trash and pollution from recreational centres and boats are to blame for this.

Disturbance of Wildlife: The presence of recreational activities and visitors has the potential to disrupt the natural behaviours and reproductive cycles of animal species that inhabit salt marshes. Stress caused by human activity, such as noise and movement, may reduce the reproductive success of animals that are especially vulnerable to human influence. The delicate equilibrium of a wetland's feeding and resting zones could be upset by activities like boating, fishing, and watching birds. This might potentially disrupt the ecosystem of the marsh, which is now in a precarious state due to a lack of biodiversity.

Introduction of Invasive Species: There is a higher chance that invasive species will be introduced into salt marshes because of the presence of leisure activities and visitors. Boats, equipment, and vehicles that have not been properly cleaned and inspected can introduce non-native plant and animal species, illnesses, and invasive plant species into the ecology of wetlands. The competitive advantage that these invasive species have over native species and the damaging affects that they have on natural habitats suggest that their introduction into a wetland environment may have a negative influence on the total biodiversity of that ecosystem.

Soil and Water Pollution: People's trash, sewage, and chemical runoff are only a few of the pollutants that can be released into the environment as a result of tourism and other forms of entertainment. Damage to wetland ecosystems and a decline in water quality may result from improper waste management practices including dumping plastics, food waste, and other trash. Human activities, including operating motorised boats, using gasoline, and applying chemical-based goods, all contribute to the entrance of pollutants into the water. Because of this, it's possible that aquatic habitats and the animals that live in them would suffer.

Erosion and Sedimentation: Inadequately managed recreation and tourism activities have the potential to exacerbate erosion and sedimentation processes already present in salt marshes. Erosion rates can occur as a result of human activity that disturbs the soil and plants, such as increased foot traffic, illegal camping, and the construction of infrastructure. The accumulation of silt can destroy aquatic habitats, reduce water clarity, and endanger the survival of plant and animal life that depends on the bottom of water bodies.

Cumulative Impacts: Salt marshes may come under increased pressure from a rise in the number of people visiting for recreational purposes and the number of visitors. Over time, environmental problems such as habitat loss, reduced biodiversity, and increased pollution can accumulate in areas that are repeatedly used. Wetland ecosystems' capacity to maintain ecological integrity and resilience may be damaged due to the cumulative nature of these repercussions.

2.6.13 Infrastructure Development

The destruction of habitat, the fragmentation of existing habitat, and the modification of hydrological processes can arise from the development of highways, bridges, and other types of infrastructure in or near salt marshes. Wetland ecosystems and the relationships within them are vulnerable to disruption from human development projects. More development increases the risk of harm to salt marshes and the delicate organisms that live there. The destruction of salt marshes can be attributed to the increased construction of roads, buildings, ports, and industrial facilities that accompany rising human populations and expanding economies. Roads, buildings, ports, and manufacturing facilities are all examples of infrastructure.

Habitat Loss and Fragmentation: Habitat loss, such as in a salt marsh, is an unexpected consequence of clearing land for infrastructure construction. Vegetation in wetlands may be lost or damaged, and this may include potentially vital salt-tolerant plant species. As a result, many species have witnessed declines in their breeding and feeding grounds as well as other natural settings. The loss of biodiversity, the disturbance of biological processes, and the limitation of the migratory range of species are all exacerbated by the increasing fragmentation of wetland habitats.

Alteration of Hydrological Regimes: The disruption of the salt marsh's normal hydrological cycles is caused by the building of infrastructure there. Water flow patterns can be altered, waterways redirected, and ecological systems that normally store and replenish water supplies are disrupted by the construction of dams, canals, and drainage systems. The actions of humans can have such results. Changes in flood patterns and water availability are two consequences that may result from these alterations to the hydrological functioning of wetland ecosystems. Changing the salinity of the soil and the distribution and abundance of aquatic species are just two examples of how human activity can have a major effect on the hydrology of a wetland.

Water Pollution: The introduction of new infrastructure in salt marshes may exacerbate the problem of water contamination. Various pollutants can be carried into wetland water bodies by runoff from cities, factories, and building sites. Sediments, chemicals, heavy metals, and nutrients are all examples of these contaminants. Water pollution can destroy aquatic life, reduce water quality, and disrupt the delicate ecological balance that sustains wetland ecosystems.

Barrier to Species Movement: Physical obstacles, such as roads, highways, and barriers, can obstruct the passage of species inside and between salt marshes. Populations may become increasingly fragmented, genetic diversity may decline, and the species' sensitivity to environmental changes may grow if these constraints impede migratory patterns and access to key habitats. Species that have adapted to life in wetland habitats may be more resilient to the effects of climate change and other environmental shifts because of the isolation they experience there.

Disruption of Ecosystem Services: Saline wetland areas are helpful to the ecosystems they are a part of because of the water filtration, stormwater management, and carbon storage they provide. Wetland areas' capacity to continue offering the advantages they now do may be diminished if new infrastructure construction has an effect on these functions. Water infiltration could be reduced, for instance, by installing impermeable surfaces and drainage systems. This would make flooding more likely and impair the wetland's ability to act as a flood control mechanism in a natural way. The ability of wetlands to filter out pollutants and keep water quality high may be jeopardised if vegetation in wetlands is wiped away to make space for infrastructure development.

Cumulative Impacts: Multiple infrastructure projects being carried out in close proximity to one another in the same region may exacerbate threats to the ecosystems of saline wetland areas. More far-reaching and long-lasting effects are possible when numerous related activities are implemented simultaneously or in close proximity to one another. Saline wetland ecosystems are under threat from a number of factors, including an increase in habitat loss, changed hydrological regimes, ecosystem fragmentation, and cumulative pollution.

2.6.14 Erosion and Sedimentation

Sedimentation can become severe in salt marshes due to wind and water erosion from nearby areas and changes in hydrological patterns. Excessive sedimentation can bury benthic habitats, alter water quality, and impede the reproductive success of wetland species. Erosion and sedimentation are two of the biggest threats to saline wetland habitats, both of which can have serious consequences for the wetland's biological health and functionality. Because of their unique characteristics and delicate ecosystems, saline wetland areas are particularly susceptible to erosion and sedimentation. The effects of erosion and sedimentation, two major threats to salt marshes, will be discussed here.

Loss of Wetland Vegetation: Erosion and sedimentation pose a threat to wetland vegetation, which in turn threatens the wetland's ability to protect the soil and provide a habitat for a diverse array of organisms. Erosion can wash away the topsoil, taking with it potentially vital nutrients and organic matter, leaving behind unusable land. Wetland plant diversity and vital habitat for animals and plants that rely on these locations for survival are both threatened as a result of this decline in wetland vegetation cover.

Disruption of Hydrological Balance: The delicate hydrological balance that salt marshes attempt to preserve can be disrupted by both erosion and sedimentation. If sedimentation is allowed to build up and choke water channels, streams, and ponds in the wetland, it can disturb the normal flow patterns of water and the wetland's capacity to hold water. Changes in water levels, reduced water flow, increased floods, or localised drying out could result from this disturbance. The ecological function of the wetland ecosystem may be negatively impacted as a direct result of these changes.

Impacts on Water Quality: Sediments carried by erosion can contaminate salt marshes with things like toxic metals, weed killers, and extra nutrients. Among the many ways these contaminants can degrade water quality is by making it more turbid, decreasing the amount of light that can enter the water, and changing the water's chemical composition. Water oxygen levels may have decreased due to increased silt loads. The result is a shift in the delicate balance that exists within the wetland environment, which in turn increases the risk of extinction for aquatic species.

Loss of Habitat Complexity: Erosion and sedimentation are two processes that might reduce wetland biodiversity. This is especially problematic in a salt marsh. Microtopographic features like channels, lakes, and depressions give the wide variety of plant and animal life found in wetland ecosystems a wide variety of places to live. The environment loses part of its complexity when sediments accumulate and fill in these unique features. This is leading to the loss of specialised niches, which threatens the wetland's biodiversity.

Impact on Fish and Aquatic Species: The fish and other aquatic species that call salt marshes home are especially at risk from the harmful consequences of processes like erosion and sedimentation. Reduced reproductive success and population declines can result from sediments burying and suffocating fish eggs, fish larvae, and other species that reside at the bottom of bodies of water. Water bodies like lakes and rivers have sediments at their bases. Fish that rely on sight may have their food and reproductive habits disrupted by increased turbidity.

Additionally, it threatens to destabilise the intricate web of species-specific interactions that are essential to the health of the wetland ecosystem as a whole.

Soil Fertility and Nutrient Cycling: Erosion can progressively remove the rich topsoil layer in salty wetland ecosystems, which is essential for plant growth because of the minerals it contains. This may be harmful to the environment. The soil that remains after silt is removed may become less fertile and nutrient-rich as a result. This will have an effect on the development and production of wetland vegetation. In turn, this could affect the productivity and health of the ecosystem by disrupting internal nutrient cycling processes inside the marsh.

Soil Stabilisation: Soil stabilisation measures like revegetation, erosion control structures, and bioengineering can help prevent erosion and stabilise the soil in salty wetland environments. Outside of salt marshes, these measures can also assist stabilise the soil. In order to reduce the risk of soil erosion and promote habitat recovery, planting native plants with broad root systems may be an effective solution.

Sediment Trapping and Retention: Sediment basins, retention ponds, and sediment traps can be constructed upstream of a wetland to slow or stop the movement of sediments. Sediment can be captured and stored with the help of these features. These structures may aid in reducing sedimentation rates inside the wetland, protecting downstream areas from silt overload.

Land Management Practices: Sustainable land management practices can be useful in helping to protect salt marsh shorelines from erosion and sedimentation. Implementing methods like contour ploughing, cover cropping, and decreased tillage can lessen soil erosion and sediment flow into the wetland.

Buffer Zones: Putting up buffer zones around the edges of salt marshes and wetlands can prevent silt from washing in from neighbouring land areas. Native vegetation can be used to create these buffer zones, which does double duty by both filtering and absorbing sediments and mitigating their harmful impacts on the wetland ecology.

Education and Awareness: Educating and creating awareness among landowners, local communities, and stakeholders about the importance of managing erosion and sedimentation in salt marshes can be very effective in developing a sense of responsibility and encouraging the adoption of optimal management measures.

2.6.15 Lack of Awareness and Conservation Efforts

Lack of understanding of the biological value of salt marshes might lead to insufficient protection measures being taken. The lack of conservation efforts threatens the long-term survival of these ecosystems by exacerbating the issues they already face. Due to a lack of public awareness and conservation efforts, saline wetland ecosystems around the world are at grave risk. Even though they are unique and serve a crucial ecological function, saline wetlands are often overlooked and undervalued in favour of other habitats. Consequences of failing to recognise the importance of conserving salty wetlands and making no effort to do so are discussed below.

Limited Understanding of Ecosystem Values: As long as people are oblivious to the ecological responsibilities and values provided by salty wetland habitats, it is probable that they will continue to be underestimated and ignored. These ecosystems perform essential tasks like water purification, carbon sequestration, and providing a home for a diverse array of plant and animal life. Inadequate conservation measures and inadequate protection of these vulnerable habitats can arise from a lack of information on these characteristics.

Addressing the lack of awareness and conservation efforts for saline wetlands requires concerted actions on multiple fronts:

Education and Awareness: The importance of salt marshes and the ecological services they perform must be brought to the attention of the general people. Education, volunteerism, and public campaigns can highlight the value of these ecosystems and inspire people to work toward protecting them.

Policy and Legislation: Saline wetland regions must be protected and preserved through the creation and enforcement of severe policies and laws. Establishing protected areas, developing conservation policies, and enforcing legislation to curb habitat loss, pollution, and unsustainable development practices are all part of this effort.

Research and Monitoring: The state, possible threats, and conservation needs of salt marshes can be better understood with the help of scientific studies and monitoring programmes. This information can be used to inform management decisions and prioritise conservation efforts.

Stakeholder Engagement: The local community, landowners, conservation organisations, and government authorities must all be involved for salty wetland conservation efforts to be successful. By working together, we can cultivate a sense of mutual accountability, spread the word about the need for sustainable land management, and inspire more people to get involved in conservation initiatives.

Restoration and Conservation Projects: Putting into action restoration programmes that focus on rebuilding damaged salty wetland areas can help reverse habitat loss and degradation. Restoring the pre-damage hydrology, stopping the spread of invasive species, and promoting the growth of native plants are all crucial to a successful restoration project.

International Cooperation: The protection of salt marshes requires the involvement of multiple governments and international organisations. When the wetland area spans international borders, this is especially true. Sharing knowledge, best management practices, and resources has the potential to strengthen conservation efforts and guarantee the long-term security of these crucial ecosystems.

References

Chauhan M, Gopal B (2016) Saline wetlands of the arid zone of Western India. The Wetland Book
Hammer UT (1986) Saline lake ecosystems of the world, vol 59. Springer Science & Business Media
Weller MW (1999) Wetland birds: habitat resources and conservation implications. Cambridge University Press
Williams WD (1996) The largest, highest and lowest lakes of the world: saline lakes. Internationale Vereinigung für theoretische und angewandte Limnologie: Verhandlungen 26(1):61–79
Zheng M (1997) An introduction to saline Lakes on the Qinghai—Tibet plateau, vol 76. Springer Science & Business Media

Salinology

3

Abstract

This chapter of this book, titled "Salinology", invites readers on an intriguing exploration of the world of salinity—the comprehensive study of the physical, chemical, and biological aspects of saline environments. Often overshadowed by freshwater sciences, salinology offers indispensable insights into the distinct characteristics and dynamics of salinised ecosystems. The chapter unfolds in three sections, each delving into a different facet of salinology. The first section, "Physical Properties", delves into the unique physical attributes that define saline environments, such as salt flats and saline wetlands, crucial for understanding their ecological processes and challenges. Following this, the focus shifts to the "Chemical Properties" in the second section, elucidating the chemical composition of saline ecosystems and its impacts on both abiotic and biotic components. The intricate relationship between chemical properties and the thriving life within saline environments underscores the remarkable adaptations to extreme conditions. Lastly, "Biological Properties" takes centre stage, offering a glimpse into the diverse array of organisms, from halophytes to halophiles, and their strategies for thriving amidst high salinity. This chapter, "Salinology", provides a holistic understanding of saline environments, unravelling the interplay of physical, chemical, and biological properties that shape their unique dynamics and ecological significance. Beyond scientific curiosity, salinology offers invaluable insights into how these environments function and respond to environmental changes, including those influenced by climate change.

Keywords

Biological Properties · Chemical Properties · Physical Properties · Salinology

L. K. Sharma, R. Naik, *Conservation of Saline Wetland Ecosystems*,
https://doi.org/10.1007/978-981-97-5069-6_3

Salinology is the scientific study of saltwater ecosystems. The study of saltwater environments and all the processes that occur in them, including their physical, chemical, and biological characteristics. The unique features and dynamics of saline systems are studied by the field of salinology, which draws from geology, hydrology, chemistry, biology, and ecology.

Salinology is the study of saline waters and their properties, including their salinity, the quantity of dissolved salt they contain, the minerals that make up these waters, and the physical processes that create saline environments details are given in Fig. 3.1. The study looks into the causes of salt deposits, the evolution of salt pans, and the consequences of saltwater intrusion into freshwater aquifers, among other interactions between salty waters and the local geological formations. The research focuses on finding out where all that salt came from.

The chemical aspects of salinology include the investigation of the elements and compounds that make up saline waters, as well as the chemical processes and reactions that occur in such places. Measurements include pH, ion concentration, nutrient dynamics, and the impact of salinity on the solubility of a wide range of compounds (Mold 2012).

In addition, salinologists explore the ecological communities and adaptations of organisms that thrive in salty settings as part of their study of the field's biological properties. The study of saline ecosystems, including wetlands, salt marshes, mangrove forests, and others, is crucial to understanding the specific microbial communities that thrive under these conditions. Plants that can survive in extremely salty conditions are called halophytes.

The assessment of the effects of salinity on water quality and ecological health, as well as the management and conservation of saline ecosystems, the responsible use of saline resources, the prevention of saltwater intrusion into freshwater

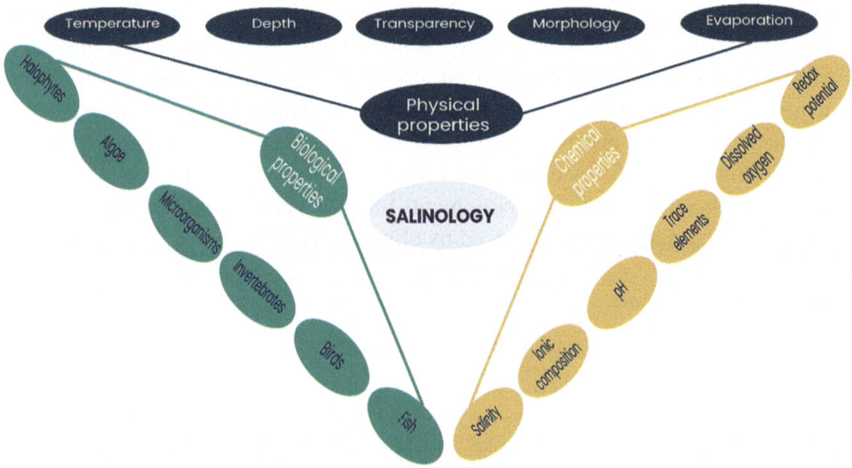

Fig. 3.1 Elements of salinology

systems, and so on, all necessitate an understanding of salinology. These are only some of the various applications of salinology knowledge.

Understanding the physical, chemical, and biological processes that occur in saline settings is essential for guiding efforts to protect and sustainably manage these unique and, at times, fragile ecosystems, and this is where salinology comes in.

3.1 Physical Properties

The presence of high concentrations of dissolved salts in water is a defining feature of saline lakes. The physical properties of saline lakes can vary depending on a number of factors, including the salt level, temperature, depth, and location. Here are some of the most notable physical features of saltwater lakes:

3.1.1 Temperature

The geology and climate of their particular sites can cause a wide range of temperature changes in saline lakes. Salty lakes may have higher surface temperatures in countries with warmer climates, whereas in parts with cooler climates, these lakes may freeze over or be covered by ice during the winter. The density of the water, the stratification of the water, and the distribution of aquatic organisms are all physical and biological phenomena that could be influenced by changes in the temperature of salt lakes. Location, climate, time of year, and lake depth are only some of the factors that might impact the temperature of saline lakes. A saline lake's temperature can be affected by the sun's rays, the air's temperature, the lake's circulation, and the surrounding environment. The following are some broad temperature-related properties of salt lakes (Oren et al. 2009):

Seasonal Variation: Changes in water temperature occur often in saltwater lakes throughout the year. The surface waters of salt lakes tend to be warmer in the warmer seasons like summer, but they might be colder in the colder seasons like winter. The temperature swings may be more extreme in regions with clearly defined seasonal boundaries.

Thermal Stratification: Larger-than-normal and deeper-than-average saline lakes are more likely to exhibit thermal stratification or the layering of water at various temperatures. The epilimnion, the highest layer, is warmer in the summer than the hypolimnion, the below layer, which remains cooler throughout the year. The transitional zone between the epilimnion and the hypolimnion is called the metalimnion, also known as the thermocline.

Mixing Events: In saline lakes, thermal stratification can be disturbed by mixing events. Strong winds, storms, or variations in air temperature can cause the lake's surface and deeper waters to mix, resulting in a more uniform temperature profile around the lake. Mixing occurs more frequently in shallower saltwater lakes than in deeper lakes.

Influence of Climate: Saline lakes in areas with a lower average yearly temperature or higher latitudes often have a lower average temperature even when the weather is warmer. Salt lakes, on the other hand, tend to have higher average temperatures when they are situated in warmer climates or at lower latitudes.

Influence of Depth: A salt lake's temperature may change depending on how deep it is. Because they are shielded from the effects of the air and sunlight, deeper waters are often cooler than the top ones. The temperature of the air has a greater impact on the surface waters. The depth of a salt lake, along with other environmental factors, can affect the temperature differential that exists inside the lake's water column.

3.1.2 Depth

The depth of saline lakes varies greatly and may be anything from extremely shallow to quite deep. Some salt lakes may be quite shallow since they are located in dry or semiarid regions that receive little precipitation. Lakes in regions of tectonic or volcanic activity often have deeper basins or have been susceptible to subsidence, making some of these lakes quite deep. Different salt lakes can have widely varying depths due to factors such as their location, geological processes, climate, and historical developments. Depending on its size, a saline lake may be a shallow body of water or a deep basin. Here is a rundown of several broad characteristics that tend to increase with lake depth in saline environments (Perillo et al. 2018):

Shallow Saline Lakes: Shallow saline lakes are lakes that have a high salinity yet a low depth. They are often shallower than 16 feet (5 m), with some instances far shallower. Evaporation at the lakes' surfaces is the primary factor influencing water levels, but changes in precipitation, evaporation, and the dynamics of inflow and outflow can all have significant effects. Shallow, salty lakes are more common in dry and semiarid regions. Comparable bodies of water include Lake Urmia in Iran and the Salton Sea in California, both of which have average depths in the range of 5–7 m (16–23 feet).

Intermediate Saline Lakes: The typical depth range for lakes that are considered intermediate in saltiness is 5–30 m (16 to 98 feet). These lakes may exhibit significant thermal stratification, which means that there are likely to be distinct layers of varied temperatures at different depths during different periods of the year. Lake Eyre in Australia and the Dead Sea are two examples of lakes that fall under the "moderate saltiness" category. In comparison to the Dead Sea's maximum depth of about 304 m (997 feet), Australia's Lake Eyre has a maximum depth of about 6 m (20 feet), however during dry periods, it can be mostly dry.

Deep Saline Lakes: When it comes to lakes, those with a high salinity tend to be quite deep, sometimes reaching hundreds of metres. These lakes are often associated with tectonic activity or the existence of deep basins. The Caspian Sea is the world's deepest saline lake at a maximum depth of about 1025 m and is the world's largest saltwater lake. It's also the longest inhabited body of water on Earth (3363 feet). Another example of a deep lake is Russia's Lake Baikal, which has a maximum depth of about 1642 m despite not being a salt lake (5387 feet).

3.1.3 Transparency

Biological activity, the amount of dissolved salts in the water, and the presence of suspended particles can all alter the clarity of the water in salt lakes. The high salinity levels found in salt lakes assist in keeping the water clear by inhibiting the growth of algae and other forms of floating debris. However, the water in some salt lakes might be murky or turbid due to the presence of mineral particles or organic debris. Transparency describes the degree to which the water in salt lakes is clear. The presence of suspended particles, dissolved chemicals, and biological activity all have an impact on the water's quality. Conditions such as these affect how much of the lake bottom may be visible through the water in saline lakes. Some points to remember about the clarity of saltwater lakes are as follows:

Water Clarity: Clear saline lakes allow light to penetrate further into the water column than lakes with cloudier water. It is commonly believed that low levels of suspended particles and dissolved compounds are to blame for this clarity. This directly causes the water to seem transparent, allowing for the observation of objects at greater depths.

Suspended Particles: The purity and transparency of water deteriorate as suspended particles like silt, algae, and organic matter accumulate in it. More suspended particles in a saltwater lake could block some of the light from reaching the lake's surface, decreasing the lake's clarity.

Dissolved Substances: Saline lakes typically have cloudier water due to the presence of dissolved salts and other contaminants. A high concentration of dissolved compounds, such as those found in hypersaline lakes, can make the water opaque and dense. This is especially noticeable in the Dead Sea and other salty bodies of water, where the salt can make the water appear murkier.

Algal Blooms: Algal blooms, which occur when some kinds of algae experience rapid growth in the water, are more common in lakes with a high salinity. Algal blooms can reduce the water's clarity and transparency, giving it a greenish or muddy appearance. Temperature, light, and nutrient availability are only a few of the factors that might trigger these blooms.

Human Influence: Human activities such as pollution, fertiliser runoff, and sedimentation can significantly reduce salt lake opacity. Pollution and the introduction of an excessive amount of nutrients can both promote algal bloom and reduce water clarity. Erosion and other land use changes can lead to sedimentation, which can introduce small particles into the water and reduce its transparency.

3.1.4 Lake Morphology

Uneven shorelines, bays, inlets, and islands are just some of the morphological features that can be found in saline lakes. The physical characteristics of saltwater lakes are influenced by a variety of geological processes, including tectonic activity, sedimentation, and erosion. Hydrodynamics, water circulation, and ecological features may all be affected by a salt lake's shape and size. Morphology refers to the overall structure, make-up, and properties of saltwater lakes. The morphology of saline lakes can exhibit a wide variety of morphological properties due to influences from geology, tectonic activity, climate, and hydrological processes. Common lake morphologies found in saltwater lakes include the following (Scott et al. 2014):

Simple Basin: One characteristic of a basin with simple morphology is that its outline is either broadly round or elliptical. In most cases, depressions or basins that are later inhabited by these salt lakes are formed as a result of tectonic action, subsidence, or volcanic activity. One such feature is the prevalence of smaller playas or salt pans in arid regions.

Complex Basin: The basin morphology of a saline lake is often complex, taking on an irregular or more twisted appearance. These lakes often have shorelines that have complex shapes like inlets, bays, and peninsulas. Extremely complicated basins can result from a confluence of several geological processes, such as faulting, folding, or erosion. The United States' Great Salt Lake in Utah has a complex basin shape due to its many bays and islands.

Sinks and Sinkholes: Some salt lakes form in sinks or sinkholes, depressions or cavities in the ground formed when soluble rocks like limestone dissolve and seep to the surface. The world is littered with sinks and sinkholes. While the shapes of these lakes are highly variable, cliffs and steep slopes are typically what set them apart. A "dead sea", like the Dead Sea, is a salt lake that formed inside of a sinkhole.

Lagoon: Protected from the open ocean by a chain of islands, sandbars, or coral reefs, a lagoon is a type of coastal lake. Most lagoons are quite shallow and stretch out for quite some distance. Lagoons come in all sorts of shapes and sizes, from rectangular to oval to erratic. As a result of their proximity to beaches and the processes that take place there, they are also influenced by the motion of the waves. Lagoon-style saltwater lakes are relatively rare, although examples include Lake Chilika in India and Ria Formosa in Portugal. These two lakes can both be found in Asia.

Deltaic: The existence of river deltas that extend into the body of water is indicative of the deltaic shape found in some salt lakes. These deltas form when rivers carrying sediment deposit it at the lake's water intake, where it becomes a permanent part of the landscape. As sediment accumulates, distributaries form as a natural consequence. Several river deltas, most notably the Volga Delta, contrib ute to the Caspian Sea's distinctive deltaic form, making it the largest inland saltwater lake.

Fjord-like: In places that have been subject to glaciation or geological activity, salt lakes may take on the appearance of fjords. The walls or cliffs of these lakes are extraordinarily steep, making them exceptionally long, narrow, and deep. Examples of saline lakes with the appearance of fjords and located in high-latitude or mountainous settings include the fjords of Norway and the deep coastal lakes of New Zealand.

3.1.5 Evaporation

Saline lakes are typically located in arid or semiarid zones, where precipitation is scarce and evaporation is high. The high rates of evaporation in the lake cause salts to concentrate in the water, which raises the salinity of the lake. Evaporation affects the volume and level of saltwater lakes, which can cause changes in the lakes' size and hydrological conditions. Evaporation has a significant role in the hydrological transition that occurs in salt lakes. In saline lakes, evaporation is an important mechanism that controls the water balance and the salinity levels. It is common knowledge that saline lakes are distinguished by their high salt content. Here is a summary of what you need to know about salt lake evaporation:

Evaporation Rate: Saline lakes' water evaporates more quickly than freshwater lakes because of the higher concentrations of salt in the former. Vapor pressure is reduced and boiling point is raised due to the high salt content of the water. As a result, it becomes more difficult for water molecules to escape into the atmosphere. Because of this, a greater quantity of water evaporates from the lake's surface.

Impact on Water Level: Evaporation plays a key role in the water balance of salt lakes. The water level of a lake can gradually decrease due to the loss of water that might occur due to evaporation, especially in dry or semi-arid conditions where evaporation rates may exceed precipitation rates. Because of this, the lake's water level may drop. As a result, playa deposits or salt flats may become exposed as the surface area of lakes decreases.

Salinity Concentration: A salt lake's residual water contains a higher concentration of salt as a result of evaporation. When the lake's salinity grows, it's because more of the dissolved salts that were already there have been concentrated there. This happens commonly in saline lakes with closed basins, meaning there are no outlets for the water to drain out. Therefore, salts gradually accumulate in the water over time.

Salt Crust Formation: Due to the high rate of evaporation, saline lakes often develop salt crusts and salt pans on their surfaces. Crystalline salt deposits form on the bottoms of lakes and other bodies of water from the salts that are left behind as water evaporates. Oftentimes, salt crusts add a unique aesthetic to their surroundings and serve as a focal point.

Influence on Ecosystems: Rapid evaporation and growing salinity levels in saline lakes can make life challenging for a wide range of organisms. Many aquatic species may find increasingly salty water uninhabitable, leading to a net loss of biodiversity if this trend continues. However, halophytes and halophilic bacteria are just two examples of extremophile species that have adapted to thrive in extremely salty environments.

Climate and Regional Variation: Evaporation rates in salt lakes are affected by climatic factors such as temperature, humidity, wind speed, and sun radiation. High temperatures and low relative humidity are typical in arid and semi-arid regions, where evaporation rates are consequently high. Furthermore, a lake's evaporation rate can be greatly affected by its temperature and location, leading to a wide range of evaporation patterns and rates for saltwater lakes worldwide.

3.2 Chemical Properties

Saline lakes are distinguished from other types of lakes by their unique chemical composition, which is a result of the large amounts of dissolved salts they contain. Several factors, such as local geology, the types of salts present, and hydrological processes occurring inside the lake, can influence the chemical composition of saltwater lakes. Some of the most notable chemical features of salt lakes are as follows:

3.2.1 Salinity

The salinity of a lake is a useful measure of its chemical composition. The concentration of dissolved salts in water is often expressed as a percentage or ppt (per cent). Saline lakes, in contrast to freshwater lakes, contain significantly higher concentrations of salt. Sodium chloride, or common table salt, is the most common type of salt found in saline lakes; however, other types of salts, such as magnesium, calcium, and potassium salts, may also be present. Saline lakes get their name because of the high levels of salinity that are characteristic of these ponds. Salinity is typically expressed in either ppt or as a percentage, depending on whether the dissolved salt content is being utilised as the standard unit of measurement (per cent). Salinity levels in saline lakes range from relatively mild (10–30 ppt) to extremely severe (>1000 ppt) (above 40 ppt or even higher). They are highly salty because of the small amount of freshwater that flows into them and the massive amount of water that evaporates out of them. Depending on their location, climate, hydrology, and the processes that occur within the earth's geology, saline lakes can experience tremendous shifts in their salt levels. The concentration of dissolved salts in the water is typically used as the universally accepted scale for describing salinity. Parts per thousand (ppt) or % are common units for expressing this concentration (per cent). Here are several broad classifications of saltwater lake salinity:

Hypersaline: The salinity of hypersaline lakes is substantially closer to or even higher than that of saltwater. These lakes have a salinity that varies widely, from below 50 ppt to well over 300 ppt. The Dead Sea, located between Israel and Jordan and boasting a salinity of roughly 340 parts per thousand, is one of the most famous examples of a hypersaline lake.

Very Saline: Extremely salty lakes have a high salinity, however, they may not be as salty as hypersaline lakes. Their saltiness often lies between 20 and 50 ppt. Depending on where you are in Utah, United States Great Salt Lake, salinity ranges from about 5 parts per thousand (ppt) to about 27 ppt.

Moderately Saline: Moderately salty lakes are those with salinities in the centre of the range. Their salinity typically falls between 5.0 and 20.0 ppt. The majority of the salt lakes found on every continent can be found in this group. Lake Eyre in Australia, is the continent's largest salt lake, with a salinity range of about 18 parts per thousand during dry periods and near-freshwater levels during floods. Australia is home to Lake Eyre.

Brackish: Most brackish lakes have a salinity level in the middle of the range, between 0.5 and 5 ppt. They contain both freshwater and saltwater, however, the amounts of both can vary widely. Different factors, such as the amount of freshwater that flows into a brackish lake, the amount of water that evaporates, and the influence of the tides, can cause the salinity of a lake to rise or fall over time. The Baltic Sea is an example of a brackish body of water due to its varying salinity, which is roughly 2% in the north and 10% in the south.

3.2.2 Major Ions

Numerous major ions contribute to the complex chemical composition of saltwater lakes. Common examples of anionic ions are sodium (Na^+), chloride (Cl^-), magnesium (Mg^{2+}), calcium (Ca^{2+}), potassium (K^+), sulphate (SO_4^{2-}), and carbonate (CO_3^{2-}) (CO_3^{2-}). The precise concentrations of these ions across the lake determine the lake's overall chemical signature. Saline lakes are distinguished by the high levels of dissolved salts they contain; these salts are mostly composed of different ions. The most prevalent primary ions in salty lakes are as follows:

Sodium (Na^+): Sodium is one of the most abundant ions in saline lakes. It is derived from weathering of rocks and is a dominant cation in many saline environments.

Chloride (Cl^-): Chloride ions are commonly present in high concentrations in saline lakes. They are derived from the dissolution of chloride-containing minerals and are often associated with sodium.

Sulphate (SO_4^{2-}): Sulphate ions are another major constituent of saline lakes. They can originate from the oxidation of sulphide minerals and are commonly found in association with sodium and calcium.

Calcium (Ca^{2+}): Calcium ions are often present in significant amounts in saline lakes. They can come from the dissolution of calcium-bearing minerals and are usually associated with sulphate and carbonate ions.

Magnesium (Mg^{2+}): Magnesium ions are commonly found in saline lakes and are derived from the dissolution of magnesium-containing minerals. They are usually associated with chloride, sulphate, and carbonate ions.

Carbonate (CO_3^{2-}): Carbonate ions are present in varying concentrations in saline lakes. They can be derived from the dissolution of carbonate minerals and are often associated with calcium and magnesium.

3.2.3 pH

The pH of salt lakes varies with the types of salts contained in the lake and the geological aspects of the landscape surrounding the lake. Saline lakes can display a wide range of pH values, from extremely acidic to extremely alkaline. One of the most common forms of lake, alkaline saline lakes have a pH greater than 8.0. Several factors, such as the type of salts that have dissolved and the surrounding geology, as well as the influence of biology, can affect the pH of a saltwater lake. Saline lakes can display a wide range of pH values, from extremely acidic to extremely alkaline. Some broad things to keep in mind about the pH of salt lakes are as follows:

Alkaline pH: Many salt lakes have pH readings that lean towards the alkaline side of the scale. This is usually the case because the dissolved salts contain a large percentage of alkaline chemicals like carbonates and bicarbonates. Lakes with a pH greater than 7 are typically alkaline and salty, and some lakes can reach even higher pH levels of 9 and 10. The high alkalinity could have resulted from the weathering of alkaline rocks or the creation of carbonates from evaporated or concentrated water.

Carbonate Buffering: High carbonate and bicarbonate concentrations are associated with greater pH buffering capacity in saline lakes. In other words, the presence of acidic or basic chemicals won't drastically alter the pH level. The relatively high pH level in salt lakes is maintained via the carbonate buffering process.

Influence of Evaporation: Concentration of the dissolved salts in the water and an increase in pH levels may result from evaporation, a process that occurs regularly in salt lakes. As lake water evaporates, alkaline chemicals like carbonates and bicarbonates become more concentrated, raising the pH. Carbonates and bicarbonates are examples of these alkaline materials.

Influence of Biological Activity: Aquatic plant and algal photosynthesis and respiration are two biological activities that can alter the pH of saltwater lakes. Plants and algae use up carbon dioxide during photosynthesis, increasing the pH level. However, respiration results in carbon dioxide generation, which can decrease pH levels.

Local Geology: It's also likely that the geology beneath the salt lake affects its pH levels. There are many different kinds of rocks, and the mineral content of each one can vary widely. These minerals may alter the pH and other chemical characteristics of the lake water.

3.2.4 Trace Elements

Trace element concentrations in saltwater lakes can vary widely. Iron, manganese, arsenic, selenium, and boron are all examples of metals and metalloids that fall under this category. The presence and abundance of specific trace elements in a lake are determined by geological and hydrological processes occurring in the lake's catchment region. Saline lakes may include a wide variety of trace elements; however, their concentrations are often fairly low when compared to those of the major ions. The specific trace elements that can be detected in saline lakes may vary from lake to lake based on the geological features of the lake and the region surrounding the lake. Some of the trace elements that may or may not be present in saltwater lakes include:

Lithium (Li): Lithium is a trace element commonly found in saline lakes. It occurs in low concentrations and can be derived from weathering of lithium-containing minerals in the watershed.

Boron (B): Boron is another trace element often found in saline lakes. It can originate from geological sources and is commonly associated with the evaporation of brines or the dissolution of boron-bearing minerals.

Strontium (Sr): Strontium is a trace element that can be present in saline lakes. It is derived from the weathering of strontium-containing minerals in the catchment area.

Iron (Fe): Iron is a trace element that can be found in varying concentrations in saline lakes. It can originate from both natural sources, such as iron-rich sediments and rocks, and anthropogenic sources, including mining activities or industrial discharges.

Manganese (Mn): Manganese is a trace element that can be present in saline lakes. It can be derived from both natural weathering processes and anthropogenic inputs.

Selenium (Se): Selenium is a trace element that can occur in saline lakes. It can be present in low concentrations in the water column and sediments and can originate from both natural sources and human activities.

Arsenic (As): Arsenic is a trace element that can be found in some saline lakes. It can be derived from geological sources and can also be associated with mining activities or industrial contamination.

Uranium (U): Uranium is a trace element that can be present in saline lakes, particularly in areas with uranium-rich geological formations. It can occur in low concentrations and may be associated with other radioactive elements.

3.2.5 Dissolved Oxygen (DO)

Conditions such as temperature, depth, and biological activity can affect the total amount of dissolved oxygen in a saltwater lake. Some saltwater lakes may have less dissolved oxygen than freshwater lakes due to the increased salinity and higher temperatures. This is due to the higher concentration of salt in saltwater lakes. However, this is subject to variation depending on the local climate and the presence or absence of photosynthesis-capable species. Several factors, including water temperature, salt concentration, biological activity, and physical processes that produce mixing, can affect the concentration of DO in saltwater lakes. Here are some important things to keep in mind about the dissolved oxygen in salt lakes:

Temperature: One of the most influential factors in determining the concentration of dissolved oxygen in salt lakes is the water's temperature. Oxygen dissolvability in water decreases as temperatures rise. In general, warmer-temperature lakes have less dissolved oxygen than their colder-temperature counterparts of the same sort.

Salinity: The amount of dissolved oxygen in saltwater lakes may vary depending on the salinity. In general, when seawater salinity rises, the amount of oxygen that can be dissolved in it falls. Dissolved oxygen levels are often lower in saltier lakes like the Dead Sea than they are in fresher lakes.

Mixing and Stratification: The patterns of mixing and stratification that occur in salt lakes play a significant impact in determining the dispersion of dissolved oxygen. The flow of oxygen between the surface waters and the deeper waters can be enhanced by the vertical mixing that may occur as a result of wind, currents, or thermal activities. On the other hand, stratification is possible in saline lakes. This occurs when the lake's bottom becomes covered with denser, saltier water, which threatens the lake's oxygen supply.

Biological Activity: Dissolved oxygen levels may be affected by the presence and activity of photosynthetic organisms like algae and aquatic plants. Through photosynthesis, these creatures produce oxygen throughout the day, which helps to increase atmospheric oxygen levels. However, during the night or when algae are abundant, oxygen uptake through respiration can cause dissolved oxygen levels to drop.

Decomposition and Organic Matter: The decomposition of organic matter may cause a drop in dissolved oxygen levels in saltwater lakes. When organic matter, like dead algae or other organic detritus, builds up, it can increase the pace of microbial decomposition, which in turn reduces the amount of oxygen in the environment.

Anthropogenic Factors: Humans can change the dissolved oxygen levels in salt lakes. Water pollution, fertiliser runoff, and the release of organic or inorganic chemicals can all lower oxygen levels in water, which in turn can lower water quality. An excessive intake of nutrients can trigger algal blooms, which in turn can increase the breakdown process's need for oxygen.

3.2.6 Redox Potential

The likelihood that a chemical system may undergo an oxidation or reduction reaction is represented by the system's redox potential. Redox conditions in saline lakes can vary greatly due to the interaction of various ions and organic materials. Both the availability of nutrients and the behaviour of certain chemical species in the lake are affected by the redox potential. The redox potential, also known as the oxidation-reduction potential (ORP), quantifies the propensity of a system to gain or lose electrons. The redox potential of a salt lake is crucial in defining the availability and mobility of different chemical species, as well as the overall biogeochemical processes occurring in the lake. Here is a summary of key data about the redox potential of saltwater lakes:

Oxygen Availability: Saline lake redox potential can be affected by the amount of oxygen present in the water. More oxygen in the air usually means more oxidising conditions, which in turn means a higher redox potential. Oxygenated and well-oxygenated surface waters of salt lakes tend to have higher redox potential values.

Anaerobic Conditions: Saline lakes can have both oxygen-rich and oxygen-poor regions. Many different redox processes can occur in oxygen-depleted environments, resulting in lower redox potential values. The sediments at the bottom of salt lakes, as well as other stagnant or low-flow areas, can support anaerobic conditions.

Redox Reactions: Many different chemical species take part in the many different redox reactions that take place in saline lakes. For instance, in oxidising conditions, things including organic matter, sulphides, and ferrous ions (Fe^{2+}) are more likely to be oxidised, leading to the release of electrons. "Reducing" circumstances, on the other hand, are optimal for the reduction of compounds like nitrates, sulphates, and ferric ions (Fe^{3+}).

Redox Zonation: Salt lakes exhibit a phenomenon known as vertical redox zonation. Different redox states can be found at different depths or locations within the lake, as described by this phenomenon. Temperature, oxygen levels, nutrient inputs, and microbial activity are just a few of the many factors that can alter this zonation. The redox potential gradient affects both the horizontal distribution of chemical species and the biogeochemical processes occurring in the lake.

Microbial Activity: Microorganisms play a crucial role in determining the redox potential of salt lakes. Microbes can take part in a wide range of redox reactions, which allow for the conversion of different chemical species. Bacteria, for instance, can participate in processes that alter the redox potential, such as the oxidation of organic materials, the reduction of nitrates, and the reduction of sulphate.

Anthropogenic Influence: Human activity can significantly affect the redox potential of salt lakes. Pollutants, such as fertiliser runoff or the discharge of organic and inorganic compounds, have the potential to alter the lake's redox conditions and biogeochemical processes. For instance, introducing an excessive number of nutrients might lead to oxygen depletion and the development of deteriorating conditions.

3.3 Biological Properties

Saline lakes have distinctive ecological features due to their high salinity and unique chemical composition. These characteristics shape the diversity of organisms found in different habitats and how they have adapted to their environments. Here are some of the most notable biological features of saltwater lakes:

3.3.1 Halophytes

Halophytes are plants that can survive in extremely salty environments, and you may find many of them in saline lakes. These plants can adapt to high salinity conditions because they have systems that either exclude or tolerate salt. Examples of halophytes include glassworts, pickleweeds, and various forms of salt marsh grasses. Halophytic vegetation plays an essential role in salty lake ecosystems by contributing to sediment stabilisation, animal habitat creation, and nutrient cycling. The importance of halophytic vegetation in providing habitat for species is also crucial.

3.3.2 Classification of Halophytes

Plants that have adapted to survive in highly salty conditions are called halophytes, and they can be sorted into a wide variety of categories. Plants are divided into these groups according to their salt tolerance and the strategies they have developed to survive in environments with high salt levels. The three most common ways of grouping halophytes are as follows:

3.3.2.1 Salt Tolerance Classification

(a) Obligate Halophytes: These halophytes can't make it in soil with low salt levels since they thrive in high salt environments. They've evolved specialised mechanisms—like salt glands—to handle the increased salt in their environments.
(b) Facultative Halophytes: These halophytes can thrive in both high- and low-salt conditions, making them extremely versatile. In response to changes in salt concentration in the environment, their physiology can adapt.
(c) Recretohalophytes: These halophytes can survive in extremely salty environments for short periods. They can survive in salt water for short periods, but long-term exposure to excessive salt levels will kill them.

3.3.2.2 Salt Response Classification

(a) Secretors: These halophytes actively expel excess salt through specialised salt glands on their leaves or stems. They are able to adapt to high-salinity conditions by keeping salt concentrations in their tissues to a minimum.
(b) Excluders: These halophytes are able to restrict their salt intake in two ways: through the use of root exclusion mechanisms and ion-selective membranes. They are able to avoid a buildup of salt in their bodies.

(c) Accumulators: The tissues of these halophytes can tolerate and even accumulate large amounts of salt. They are able to keep their water balance and osmotic potential by accumulating salt in vacuoles or other compartments, even when exposed to high salinity.

3.3.2.3 Life Form Classification

(a) Halophytic Trees and Shrubs: These salt-tolerant trees are called halophytes. Succulent leaves, extensive root systems, and salt resistance mechanisms are common in these plants.
(b) Halophytic Grasses and Sedges: Grass- and sedge-like halophytes thrive in salty environments. Narrow leaves, fibrous roots, and salt adaptations are typical features of these plants.
(c) Halophytic Herbaceous Plants: Some herbaceous plant species, known as halophytes, can survive in extremely salty environments. Succulent leaves, adapted root systems, or other mechanisms may help them survive in salty environments.

3.3.3 Algae

Algae such as cyanobacteria, diatoms, and green algae can all make their home in saltwater lakes. Some forms of algae have adapted over time to thrive in environments with high salt levels. Primary producers, also provide food for other organisms and aid in nutrient cycling. An outbreak of algal blooms, which can occur when there is significant algal productivity, can disturb the ecological dynamics of a lake. The extreme salinity conditions of salt lakes are ideal for a diverse range of algae. These algae play a crucial role in maintaining the health of salt lake ecosystems. Some of the most common algal species in saltwater lakes are as follows:

3.3.4 Classification of Algae of Saline Lakes

The taxonomic categorisation, morphological features, and ecological properties of the algae that live in saline lakes allow for a number of distinct groupings. The most prevalent types of algae in saltwater lakes are as follows:

Cyanobacteria (Blue-green Algae): The photosynthetic bacteria known as cyanobacteria can be found in saltwater lakes and all the world's oceans. Their ability to perform oxygenic photosynthesis sets them apart from all other known forms of life. Common genera of cyanobacteria in saltwater lakes include *Microcystis, Oscillatoria*, *Spirulina*, and *Anabaena.*

Chlorophyta (Green Algae): Green algae are a phylum of photosynthetic organisms that can thrive in environments with a wide range of salt concentrations. Green algae are ubiquitous in both marine and freshwater ecosystems. The high levels of chlorophyll in these organisms are what gives them their distinctive green colour. Green algae, including species from the genera *Dunaliella, Chlorella*, and *Botryococcus*, are commonly found in salty lakes.

Bacillariophyta (Diatoms): Diatoms are a form of algae that flourish in saltwater lakes and other marine habitats. The silica in their cell walls is in a particularly intricate structure. Their distinctive frustule architecture makes them easy to spot. The diatom genera *Nitzschia, Navicula,* and *Cyclotella* all have numerous species that thrive in saline lakes.

Euglenophyta (Euglenoids): Oftentimes, freshwater environments are the best places to find euglenoids, a type of unicellular algae. Their forms are malleable, and they feel soft to the touch. However, you might be surprised to learn that you can find euglenoid species in saltwater lakes! *Euglena* and *Phacus* are two euglenoid taxa that are commonly found in saltwater lakes.

Xanthophyta (Yellow-green Algae): The presence of pigments like fucoxanthin gives the Xanthophyta class of algae its distinctive yellowish-green colour. Although xanthophytes are typically found in freshwater settings, some species are able to survive and even thrive in saltwater conditions. *Vaucheria* and *Tribonema* are two genera that could thrive in saline waters.

Rhodophyta (Red Algae): Red algae are a type of algae that can be found in both marine and freshwater environments, with the former being more common. Their bright red colour, caused by pigments like phycoerythrin, helps them stand out among other forms of life. Some red algae, possibly including *Porphyra* and *Bangia* species, have been found in salty lakes.

3.3.5 Microorganisms

Microscopic creatures, including bacteria, archaea, and eukaryotic species, can be found in saline lakes. There is a group of bacteria known as halophiles that have adapted to survive and even thrive in very salty conditions. They play an important role in the biogeochemical processes that occur in ecosystems that include salty lakes due to the unique adaptations that allow them to survive in high-salinity settings. Many different kinds of microorganisms can be found in salty lakes because they have adapted to live in the harsh environments. The ecosystems that thrive in salt lakes depend on these bacteria to carry out vital tasks. Some typical microbial communities in saltwater lakes include the following.

Bacteria: Numerous different kinds of bacteria can be found in abundance in saltwater lakes. They include both aerobic and anaerobic species and contribute to numerous biogeochemical processes. While certain bacteria can break down organic materials and aid in nutrient cycling, others may play a role in sulphur or nitrogen cycling. Examples of the kinds of bacteria that can be found in saline lakes include the genera *Halomonas, Salinibacter,* and *Halobacteriovorax*.

Archaea: Archaea are a type of microorganism that can survive in extremely hostile environments like highly salty lakes. Since halophilic archaea flourish in salty settings, they are by far the most abundant form of archaea in saline lakes. For instance, halobacteria are well-known as a form of haloarchaea that can thrive in excessively salty environments. Due to their pigmentation, they might sometimes be responsible for the unique hue of saltwater lakes.

Algae: Algae, both cyanobacteria and eukaryotic, are the primary producers of salty lakes. Photosynthesis, which they are responsible for, is crucial to the circulation of energy and the recycling of nutrients in these systems. Microcystis and Oscillatoria are two types of cyanobacteria that can develop dense blooms in saline lakes under the correct conditions. Primary producers in the environment include eukaryotic algae, such as green algae (like *Dunaliella*), and diatoms.

Protozoa: Single-celled eukaryotic protozoa are commonly found in salty lakes and are known as protozoa. They play crucial roles in both the microbial food web and the environmental nutrient cycle. Some ciliates and flagellates, for example, help transfer energy and nutrients up the food chain by eating bacteria and algae.

Viruses: Numerous viruses live in saline lakes and contribute to the lakes' ability to regulate microbial populations and recycle nutrients. In saline lake environments, bacteriophages, which are viruses that infect bacteria, may alter the composition and function of bacterial communities and the processes by which nutrients are recycled.

3.3.6 Invertebrates

Many different kinds of invertebrates can be found in salty lakes because they have adapted to the harsh environment. There are a variety of animals that thrive in salty environments, such as brine shrimp, brine flies, copepods, and halophilic crabs. These invertebrates play a crucial role in the ecosystem as they supply nutrition to higher creatures and aid in the recycling of waste products. Numerous kinds of invertebrates that have adapted to living in extremely salty environments can be found in saline lakes. Lakes are a part of these settings. These invertebrates perform crucial functions in salt lake ecosystems, which greatly improves the health of these environments. Some of the most prevalent orders of invertebrates in saltwater lakes are as follows:

Invertebrates found in saline lakes can be classified into many distinct taxonomic groups based on the characteristics they share and the evolutionary relationships they have. Some of the most common classes of invertebrates found in saltwater lakes are as follows:

Crustaceans: The crustaceans are a group of invertebrates distinguished by their articulated appendages and hard exoskeleton. There is a huge range in the size and form of crustaceans. Brine shrimp (*Artemia*), fairy shrimp (*Anostraca*), copepods, and ostracods are only a few of the many taxonomic forms that can be found in salty lakes.

Insects: Saline lakes, especially those adapted to high salinity, are home to a vast and varied community of insects. Insects of many different kinds make their homes in saline lakes. Beetles, mosquitoes, midges, and many other types of aquatic insects are only a few of the many taxa that make up this clade.

Molluscs: Molluscs are an extremely diversified collection of invertebrates with a soft interior protected by a tough exterior. Molluscs are ubiquitous, populating every marine and freshwater habitat on the planet. Salinity in lakes can support a variety of snails and other molluscs. They play crucial roles in the nutrient cycle and the decomposition process.

Annelids: Annelids are a class of segmented worms found in both freshwater and saltwater ecosystems. These organisms play an important role in nutrient cycling and sediment dynamics as either oligochaetes (which contain tubificid worms) or polychaetes.

Rotifers: Rotifers are tiny multicellular creatures that thrive in freshwater environments but can also tolerate somewhat salty conditions. They serve important roles as grazers and make significant contributions to the saline lake food chain.

Nematodes: The microscopic nematodes, often known as roundworms, can be found in a wide range of habitats, including brackish and salty lakes. They are crucial members of the decomposition team and make important contributions to the systems that recycle nutrients.

Other Invertebrates: Many different kinds of invertebrates, including water mites, flatworms, and tardigrades, can be found in saltwater lakes. The high salinity of the lakes likely presents no problem for these invertebrates since they've evolved specialised adaptations.

3.3.7 Birds

Waterfowl and shorebirds, among others, rely heavily on salty lakes for their existence. Some migratory birds use these places as breeding and feeding grounds, specifically those that specialise in eating brine shrimp and other invertebrates that thrive in saline conditions. They also construct their nests in these regions. Saline lakes can be vital resting spots for migrating birds, contributing to the safety of these species and the conservation of biodiversity. Numerous bird species make their home in saline lakes because of the variety of food sources and unique environment they provide. The following are examples of some of the more widespread bird species found in and around salt lakes:

Birds found in saline lakes can be classified into a wide variety of taxonomic categories due to their unique characteristics and shared evolutionary ancestry. Here is a list of some of the more common bird families that frequent areas around saltwater lakes.

Phoenicopteridae (Flamingos): Large, elongated birds with curved bills characterise flamingos. They have evolved to consume algae and crustaceans in salty lake conditions, and their bright pink plumage is a trademark. The genus and species in question are *Phoenicopteridae.*

Recurvirostridae (Avocets and Stilts): Wading birds like avocets and stilts have long, slender legs and recognisable beaks. They feed on small invertebrates and insects in shallow water and are members of the family Recurvirostridae, which is commonly found in salty lake settings.

Charadriidae (Plovers): Plovers range in size from little to medium, having short beaks and an obvious erect posture. They are members of the family Charadriidae and can be found in saltwater lakes where they eat the insects, crustaceans, and other minute invertebrates that they find there.

Scolopacidae (Sandpipers and Snipes): The sandpiper and the snipe, both members of the family Scolopacidae, are small- to medium-sized wading birds. Many species of these birds travel considerable distances during their annual migrations. During their migration, some sandpipers and snipes stop by salty lakes to feed on small invertebrates and insects that they find in the mudflats and shallows.

Anatidae (Ducks, Geese, and Swans): The Anatidae family, which includes ducks, geese, and swans, is highly represented in the areas surrounding saltwater lakes. Saline lakes may be used by these animals for reproducing, resting, and feeding, as they have evolved to a wide range of aquatic environments. Many members of this family inhabit salty lake ecosystems.

Laridae (Gulls and Terns): Laridae, which includes gulls and terns, are birds that range in size from middleweight to huge. Coastal locations are a common habitat for them, and they have been known to forage in salty lakes. While gulls are commonly associated with scavenging, terns are renowned for their fishing prowess.

Pelecanidae (Pelicans): Pelicans are huge, ponderous birds with massive throat pouches and long, snouts. The Pelecanidae family is closely linked to salty lakes with thriving fish populations. To catch fish, pelicans use the pouches on their bills.

Ardeidae (Herons and Egrets): The Ardeidae family includes the long-legged wading birds known as herons and egrets. Some members of this family inhabit saltwater lakes, where they hunt for fish, amphibians, and insects.

Accipitridae (Hawks, Eagles, and Kites): Birds of prey from the family Accipitridae, including hawks, eagles, and kites, are frequently spotted in the area around saltwater lakes. Small mammals, birds, and fish are also potential prey for these raptors in the saline lake habitat.

3.3.8 Fish

While freshwater lakes often have a wider range of fish species and larger populations, certain saline lakes do support fish populations. Several indigenous fish species can only be found in saline lakes because of their unique adaptations to the high salt content of these bodies of water. Because of their evolutionary ties and shared properties, the fish that inhabit saltwater lakes can be classified into a plethora of taxonomic families. Here are some of the many types of fish and their respective habitats that may be found in saltwater lakes:

Cyprinodontidae (Pupfishes): Pupfishes, which belong to the family Cyprinodontidae, are notable for their tolerance of high salinity. They are intelligent tiny fish that have adapted well to many salty lake environments. There are several species of pupfish found in the United States, including the Death Valley pupfish, Salton Sea pupfish, and Devil's Hole pupfish.

Fundulidae (Killifishes): Killifish, which belong to the family Fundulidae, can be found in both freshwater and saltwater environments. Some members of this family have evolved to thrive in saltwater environments, such as lakes. Killifish, despite their small size, are widely recognised for their brilliant hues.

Cichlidae (Cichlids): The cichlid family is large and varied, with members living in freshwater, brackish, and saltwater lakes. Cichlidae is a family of fish that includes several species that have developed salinity tolerance and can live in saltwater lakes.

Tetraodontidae (Pufferfishes): Pufferfishes belong to the family Tetraodontidae and are known for their ability to inflate their bodies when threatened. While not exclusively found in saline lakes, certain species of pufferfish have adapted to tolerate the high salinity conditions of some saline lake habitats.

Poeciliidae (Livebearers): Fish in the family Poeciliidae are known as livebearers because they deliver their young alive rather than depositing eggs. Some members of this family thrive in extremely salty environments and are hence commonly found in saltwater lakes. Swordtails and mollies are just two examples.

Cyprinidae (Carps and Minnows): Many different kinds of fish, such as carps and minnows, are all members of the Cyprinidae family. Some cyprinid species, while most cyprinids are freshwater specialists, can actually be found in saline lakes because they have adapted to living in brackish or salty water.

Osmeridae (Smelts): Smelts, which are members of the family Osmeridae, are a type of small, skinny fish that thrive in a wide range of salinities. There are species in this genus that thrive in saltwater environments, such as lakes.

Gobiidae (Gobies): A wide variety of environments, including saltwater lakes, are home to the goby fish family. Some members of the family Gobiidae can survive in saltwater, making them suitable residents of saltwater lakes.

References

Mold RJ (ed) (2012) Ecology of halophytes. Elsevier

Oren A, Naftz DL, Wurtsbaugh WA (2009) Saline Lakes around the world: unique systems with unique values preface. Nat Resour Environ Issues 15:I

Perillo G, Wolanski E, Cahoon DR, Hopkinson CS (eds) (2018) Coastal wetlands: an integrated ecosystem approach. Elsevier

Scott DB, Frail-Gauthier J, Mudie PJ (2014) Coastal wetlands of the world: geology, ecology, distribution and applications. Cambridge University Press

Part II

Wetlands—Past, Present and Future (Technological Point of View)

Past Technologies for Wetland Studies

4

Abstract

This chapter, titled "Past Technology", provides a historical journey through the evolution of technology utilised in environmental studies, focusing on its pivotal role in enhancing our understanding and conservation efforts. The chapter sheds light on the significance of technological advancements, particularly in enabling on-the-field studies and harnessing the power of aerial photographs for research and conservation purposes. The first section, "On-the-Field Study", delves into the methods and tools employed by scientists and conservationists before the digital era, emphasising the importance of firsthand observations and measurements in shaping early environmental research. In contrast, the second section, "Aerial Photographs", explores the transformative impact of aerial imagery on wetland research, tracing its history and elucidating its role in mapping, monitoring, and understanding environmental changes over time. Unit 2 underscores the profound influence of technology on environmental studies, from historical perspectives to contemporary applications. It highlights the continuous evolution of tools and methods, paving the way for modern techniques explored in subsequent chapters. Throughout this unit, readers are invited to appreciate the intricate interplay between technological advancements and our comprehension of the natural world, culminating in a deeper understanding of wetland ecosystems and the imperative for their conservation and management.

Keywords

Carbon Dating · Field Studies · Photographs · Technology

L. K. Sharma, R. Naik, *Conservation of Saline Wetland Ecosystems*,
https://doi.org/10.1007/978-981-97-5069-6_4

4.1 On-the-Field Study

Although these conventional mapping methods shed light on the whereabouts and nature of wetlands, they fell short in terms of precision and specificity. Wetland mapping has grown more accurate, efficient, and consistent thanks to the development of Remote Snsing (RS) technology, Geographic Information Systems (GIS), and high-resolution satellite photography.

It took a lot of time, energy, and skill to conduct a field study. However, they were crucial in laying the groundwork for our current understanding of wetland ecosystems, their ecological processes, and the conservation requirements in this area. Field investigations are still necessary for ground-truthing, validating remote sensing data, and gathering specific information about local wetland conditions, although satellite technology and remote sensing have increased our capacity for researching wetlands.

4.1.1 Types of Field Studies

Wetland ecosystems were primarily studied through field investigations until the development of satellite technology which are diagrammatically shown in Fig. 4.1. Scientists and researchers went to the wetland sites themselves to perform a variety of field studies. Before the advent of satellite technology, the following field study methods were frequently employed in wetland research:

4.1.1.1 Direct Observations

In the field, scientists would spend time directly studying wetland habitats, gathering data about the topographical qualities, flora and fauna, and hydrological processes of the ecosystem. To gain a better understanding of the dynamics of wetland ecosystems, we would study the patterns of water flow, sediment deposition, and plant succession.

Fig. 4.1 Types of field studies

4.1.1.2 Sediment Studies

Core Sampling: Taking sediment cores from the marsh allows scientists to examine the many sedimentary layers that have built up over time. These sediments may provide clues about former vegetation, in addition to data about past temperature and human impact. Pollen, seeds, and other plant remnants can be examined to help piece together the vegetation history of a marsh. The practice of core sampling is common in the study of wetlands. The goal of core sampling is to collect cores of sediment from the marsh for further study. Sediment cores are an excellent resource for learning about past temperature, sedimentation patterns, and plant life thanks to the wealth of data they preserve. The steps involved in taking a core sample are summarised here.

Core Selection: Step one entails deciding where in the marsh to take the core sediment sample. This decision is grounded in the aims of the study, prior understanding of the wetland, and data on sedimentation patterns.

Equipment and Collection: To take a core sample, unique tools like a coring device are needed. A gravity corer or piston corer is the workhorse of the coring world. These implements are made to drill down into the sediment and extract a cylindrical core sample.

Core Extraction: There is minimal disruption to the layered sediment when a coring device is put into the marsh sediment and the core is extracted continuously. The period and research questions will determine how deep the core has to be. The length of a core may be anywhere from a few millimetres to several metres.

Core Handling and Preservation: Once the core has been removed, it is placed in a core liner or plastic sleeve to prevent any further damage. Sediment layers are protected from being lost or mixed by sealing off the top and bottom of the core. The core must be properly labelled and documented so that it can be analysed in the future.

Laboratory Analysis: The sediment core is brought back to the lab where it is studied using several methods. Analysis of sediment parameters such as grain size, organic matter content, and mineral composition may involve visual inspection, subsampling, and analysis. Radiocarbon dating, pollen analysis (palynology), diatom analysis, and geochemical analysis are only some of the specialised analyses that can be conducted on subsamples taken from various layers of the core.

Interpretation: The results of the core analysis are analysed to piece together the history of the wetland. Researchers can infer past changes in vegetation dynamics, hydrology, human effects, and climate by analysing changes in sediment composition or fossilised plant remains, among other proxies (Bowman 1990).

4.1.1.3 Peat Profiling

Peat profiling includes looking at the layers of peat in a wetland from above. It's useful for tracing the development of peat, tracking the evolution of plant life, and identifying the types of wetlands that have existed over time. It's a way to investigate the layered composition of peat layers in a wetland. Peat is a layer of partially decayed plant matter that builds up in wetland areas when there is a lack of oxygen. Scientists can learn more about the process of peat production, how vegetation has evolved over time, and what kinds of wetland ecosystems are thereby creating profiles of the peat. Methods for peat profiling are outlined below.

Site Selection: Peat profiles are taken in an appropriate area of the wetland. The selection criteria include the goals of the research, ease of access, and familiarity with the wetland's past and present vegetation patterns.

Equipment: For accurate peat profiling, specialised tools like a peat corer or peat probe are needed. The goal of these instruments is to obtain a core or sample from a peat deposit by penetrating it vertically.

Peat Core Extraction: The peat deposit is cored, and a core or sample is taken, using the coring tool. The time period and research questions will determine how deep the core has to be. The length of a core may be anywhere from a few millimetres to several metres.

Core Handling: The retrieved core is then carefully handled to prevent any damage. A core liner or plastic sleeve is used to transport the core, and its top and bottom may be sealed to prevent the loss of peat layers or their mixing. The core must be properly labelled and documented so that it can be analysed in the future.

Laboratory Analysis: The vertical structure of the peat bog is studied by bringing a core sample back to the lab. Subsamples of the core can be taken at different depths to represent distinct epochs. Visual inspection, peat thickness measurement, plant remains (macrofossil) identification, and peat composition and organic matter content estimation are all possible types of analysis.

Interpretation: Peat profiling yields information that is used to piece together the timeline of peat formation and wetland evolution. It is possible to learn about past changes in vegetation, hydrology, and the environment by analysing changes in peat thickness, the types of plants that remain, and the variations in peat composition (DeLaune et al. 2020).

4.1.1.4 Radiocarbon Dating

The age of organic material in wetland sediments can be calculated using radiocarbon dating. This enables the precise dating of specific environmental modifications that occurred over the wetland's lifetime. The amount of the radioactive isotope carbon-14 ($_{14}$C) in an organic sample can be used as a proxy for the sample's age. The concentration of $_{14}$C can be used for this purpose. This method is commonly used in archaeology, anthropology, and other related subjects to determine the age of artefacts, fossils, and other types of organic remains.

As a naturally occurring element, carbon-14 (an isotope of carbon) can be discovered in trace amounts in Earth's atmosphere. Plants and animals can take in carbon-14 in two ways: through photosynthesis and the digesting of dead organisms. As long as an organism is alive, its carbon-14 content will remain relatively stable since it is constantly exchanging carbon with its environment. When a living thing dies, it stops taking in carbon-14, and the carbon-14 it already has starts to decay. This is because the organism had stopped actively consuming carbon-14.

Carbon-14 levels in the atmosphere have been steadily and exponentially declining. After waiting around 5730 years, you'll find that there's only half as much carbon-14 as there was when you started. By comparing the amount of radioactive carbon-14 to the amount of stable carbon isotopes (carbon-12 and carbon-13), scientists can calculate the age of a sample.

Before scientists can use radiocarbon dating, they need to collect a sample of the substance to be dated (bone, wood, charcoal, etc.). The sample's carbon is then converted into a form that can be measured using a wide range of techniques. Accelerator Mass Spectrometry (AMS) is a method for determining the quantity of carbon-14 in a sample by first converting the carbon in the sample into carbon dioxide gas.

The results of these analyses are typically offered in the form of a radiocarbon age, which is expressed in years before the present, also known as the number of years since 1950. Carbon-14 levels in the atmosphere were generally steady before the middle of the twentieth century, hence this time period is used as a reference by scientists. To better predict calendar ages, radiocarbon ages have recently been calibrated against other dating methods and reference materials.

Radiocarbon dating provides the most reliable results when applied to organic samples that are no more than 50,000 years old. After that length of time has gone, there is typically too little carbon-14 left to measure accurately. Potassium-argon dating and uranium-lead dating are just two of the many techniques used to determine the age of ancient artefacts (Duling 2018).

4.1.1.5 Palynology

Scientists use pollen grains and spores found in sediment cores to reconstruct historical climates and ecosystems. Pollen assemblages allow scientists to reconstruct past plant growth patterns, monitor changes in the makeup of wetland plant communities, and detect environmental disturbances.

4.1.1.6 Diatom Analysis

Diatoms are a type of microalga with a silica shell that typically live in wetlands. Ancient water quality, salinity, nutrient levels, and hydrological conditions can be deduced from the study of diatom assemblages that have been preserved in sediment cores.

4.1.1.7 Geophysical Surveys

Wetland subsurface features can be studied with geophysical techniques including Ground-Penetrating Radar (GPR) and Electromagnetic Induction (EMI). Subterranean wetland features, such as water levels and sediment layers, are elucidated by these studies.

4.1.1.8 Historical Records and Maps

Information regarding past conditions in wetland areas can be gleaned from archival materials like records, maps, and aerial pictures. This area could serve as a repository for records detailing wetland area size, land use shifts, and human impacts on wetland ecosystems.

4.1.1.9 Dendrochronology

Dendrochronology, or tree-ring dating, can be used to accurately reconstruct previous hydrological conditions in wetlands that contain tree remains that have been preserved over time. Conditions like drought, flooding, and other changes in water levels fall into this category.

4.1.2 Types of Sampling

Researchers in the field would take samples and conduct measurements inside wetland ecosystems to quantify numerous aspects. Collecting water samples for chemical analysis, measuring physical qualities including depth, temperature, pH, and turbidity, and acquiring soil samples to learn about the soil's composition and moisture content are all examples. Vegetation samples would also be collected for the purposes of species identification, biomass measurement, and a thorough analysis of the vegetation's structure.

In the past, several sampling techniques were used to gather data from populations of interest. Here are some commonly used methods:

Simple Random Sampling: For this method to work, you'll have to pick people at random from a set of possibilities inside a population. In the past, it was common practice to generate random outcomes by drawing names from a hat or selecting numbers at random from tables containing random numbers.

Systematic Sampling: Researchers use this strategy to select every nth person from a population after randomly choosing a starting point. A random starting number between 1 and 10 can be set, and then every tenth person in the population will be sampled to get the desired sample size of 100.

Stratified Sampling: This strategy uses distinguishing characteristics to divide the population into subsets or strata (e.g., age, gender, occupation). Once that is established, random samples are drawn from each stratum in proportion to the entire population. Stratified sampling is used to ensure that statistically important subgroups are adequately represented in the sample.

Cluster Sampling: First, the population is divided into subgroups (such as geographic regions or types of organisations), and then a random sample is drawn from each. After that, data is collected from everyone in the selected clusters. Cluster sampling is a useful alternative to direct individual sampling in situations where the latter is not feasible.

Convenience Sampling: Convenience sampling was widely used in the past since it was quick and easy to implement. Convenience sampling, however, is not a valid statistical technique. The process involves selecting individuals who can be reached quickly or easily. However, it may not be a representative sample of the population and hence may introduce bias.

Purposive Sampling: This method, also known as judgmental sampling, involves selecting study participants on the basis of their possession of certain attributes or expertise pertinent to the investigation. Participants are chosen at random, with no consideration given to the researchers' past information or preferences.

Visual Survey: This requires a qualitative visual assessment of the wetland region. Scientists would take notes on the different kinds of plants, the amount of water, the kinds of animals they saw, and other factors (Essington 2015).

4.1.2.1 Sampling Techniques

Transect Line Sampling: Researchers erected transect lines over the marsh and walked along them, stopping at predetermined points to make observations or collect samples. This approach allowed for the methodical investigation of wetland features in a straight line.

Plankton Nets: Microscopic organisms like phytoplankton and zooplankton were gathered using plankton nets. Towed through the water, these nets collect aquatic species for study and identification.

Sweep Nets: Using sweep nets, we collected samples of insects and other small invertebrates living in the wetland's vegetation. The net would be swept through the plants, and the critters would be collected for research.

Pitfall Traps: When collecting samples of invertebrates from the ground, pitfall traps were frequently used. These traps were underground containers with a lid and a funnel heading inside. The trap would attract insects and other small organisms, enabling scientists to study and count those populations.

Transects and Quadrats: Wetland vegetation and animal communities are routinely studied in the field using transects and quadrats for sampling and assessment. To collect data on the distribution, abundance, and variety of plant species, researchers would set up transects (linear paths) or quadrats (fixed regions). Wetland vegetation distribution and the effect of environmental conditions on plant communities can be better comprehended with the aid of this method.

4.1.2.2 Transects

In order to study a specific area, such as a marsh, transects are laid out linearly. Researchers walk along transects, stopping at regular intervals or fixed spots to collect data or make observations. The spatial distribution of vegetation, the variation of vegetation traits, and the environmental gradient within a wetland are all topics that can be investigated with the help of transects. Many kinds of transects exist, such as

Line Transects: Scientists keep track of the types and numbers of plants they find at regular intervals along a transect line (e.g., every meter). This information can be used to calculate species richness or evaluate the degree to which the transect's vegetation has changed.

Belt Transects: By doing so, they can get a more accurate representation of the area around the transect line. Within a specified width on either side of the transect line, they capture information on vegetation type, cover, or other factors. The composition of the local flora can be better understood with the help of belt transects.

4.1.2.3 Quadrats

A predetermined sample region within the wetland can be created using square or rectangular frames called quadrats. These picture frames often come in standard sizes, such as 1 × 1 m or 50 × 50 cm. Quadrats are strategically or arbitrarily placed across the study area to collect data on the diversity of plant species, the extent of plant cover, and other ecological features. A predetermined sample region within the wetland can be created using square or rectangular frames called quadrats. These picture frames often come in standard sizes, such as 1 × 1 m or 50 × 50 cm. Quadrats are strategically or arbitrarily placed across the study area to collect data on the diversity of plant species, the extent of plant cover, and other ecological features.

While quadrats are most useful for studying plant communities, they can also be used for other things including collecting soil samples. The presence, quantity, and other relevant information on plants and other species found within the quadrat are recorded by the researchers. By taking samples from a large number of quadrats dispersed throughout the wetland, they may assess the marsh's vegetation composition as a whole, among other things.

Mapping: The wetland areas they were examining would be meticulously mapped using traditional surveying techniques. This entailed manually assessing the geographical features, water bodies, vegetation zones, and boundaries of any wetlands present using surveying devices such as compasses, theodolites, and measuring tapes. These maps revealed important details on the distribution and connectivity of wetlands as well as other aspects of the local topography. Previous wetland mapping methods relied on manual observations and field surveys to determine where wetlands were located and what types of wetlands they were. These approaches consisted primarily of looking at things, counting things, and making maps on paper (Kampa and Hansen 2004).

4.2 Aerial Photographs

Aerial photography has become more important in the study of wetland ecosystems during the past few decades which are explained in Fig. 4.2. It played a crucial role in wetland mapping, monitoring, and study. Previous research on wetlands often made use of aerial photographs in the following ways:

4.2.1 Uses of Aerial Photographs for Wetland

4.2.1.1 Wetland Mapping and Classification

The use of aerial photographs allowed for the precise location and delineation of the wetlands. Expert interpreters would study the photos for clues like bodies of water, floral patterns, and geological formations. Wetland ecosystems, such as marshes and swamps, can often be identified and mapped with the help of aerial

Fig. 4.2 Uses of aerial photographs for wetlands

photographs, thanks to the distinctive colour and texture of the vegetation that grows there.

Aerial photography could help with the categorisation of the many wetland types. Visual cues, such as open water, emergent plants, or forested sections, made it simpler to distinguish between the various wetland ecosystems. Wetland habitats, their distribution, and the processes that build them up can all be better comprehended with the aid of this categorisation.

Here's an overview of how aerial photographs were utilised for wetland delineation:

Visual Interpretation: Expert interpreters would look at the aerial photos and try to pick out the features of the wetlands just by looking at them. This required searching for common indications of wetland environments, such as bodies of water, patterns of flora, and types of soil. The boundaries of wetlands would be identified by hand by drawing or marking them on the photos.

Photointerpretation Keys: To better identify and demarcate wetlands from aerial pictures, photointerpretation keys were created. Wetland areas may be distinguished from other land cover types with the help of these keys, which give a collection of characteristics including flora signatures, colour variations, texture patterns, and proximity to water bodies.

Scale and Resolution: Wetland boundaries relied heavily on the scale and resolution of aerial images. The interpreters were able to more precisely identify minor wetland features and boundaries because of the higher resolution pictures. Wetland mapping was greatly aided by the use of high-resolution, large-scale aerial images.

Comparison with Ground Truthing: In order to confirm and validate the wetland borders extrapolated from the pictures, ground truthing was frequently performed in conjunction with aerial photography. The wetland's vegetation, soil types, and hydrological features would be evaluated through field surveys to validate the accuracy of the aerial interpretation (Li and Liu 2018; Schinner et al. 2012).

4.2.1.2 Hydrological Studies

Water flow patterns, stream networks, and wetland hydrology were evaluated using aerial images. Wetland systems, stream courses, and floodplains were more easily understood with the aid of the photographs. Wetland hydrological dynamics and processes were better understood by analysing long-term changes in water levels and vegetation patterns.

Stream Network Mapping: Stream systems inside and outside of wetland zones were mapped and delineated using aerial images. Skilled interpreters would use the aerial pictures to trace the observable watercourses and make comprehensive maps of the stream system. The importance of these data to comprehending the wetland's water flow connections and spatial organisation cannot be overstated.

Watershed Boundaries: Aerial photographs were used to locate and outline the watershed's boundaries. Researchers analysed the geographic features and drainage patterns apparent in the images to determine the catchment areas responsible for the wetland's water flow. This information was crucial for understanding the hydrological processes and interactions that occur between wetlands and the surrounding environments.

Floodplain Mapping: With the aid of aerial photography, wetland and floodplain zones may be mapped and delimited. By analysing the breadth of water bodies and floodplain characteristics seen in the images, researchers were able to identify and demarcate areas vulnerable to recurrent or seasonal floods. This data was crucial for calculating flood risks and planning for floodplains.

Water Level Analysis: Aerial photographs taken at regular intervals allowed researchers to examine how marsh water levels changed over time. The dynamics of wetland hydrology might be traced back through time by comparing photographs taken at different times of the year, allowing researchers to monitor and quantify changes in water levels. This data proved useful for discerning seasonal and/or long-term shifts in the wetland's capacity to store and release water.

Hydrological Modelling: Aerial photographs have been an invaluable part of hydrological modelling studies. With the data gleaned from the photos, the researchers were able to develop and fine-tune hydrological models. Maps of floodplains and drainage systems were among the data collected. Water resource management, flood forecasting, and ecological evaluations have all benefited from these models' ability to simulate wetland hydrology and flow dynamics.

Wetland Hydrological Assessments: Aerial photographs were utilised to analyse the various hydrological features of wetlands. Wetlands, ponds, and other bodies of water would be analysed in the photos to detect depressions, which would then be measured in terms of their size, form, and distribution. Important insights into the wetland's water storage capacity, water retention period, and interactions with surface water were gleaned from these investigations.

4.2.1.3 Vegetation Studies

The use of airborne photos greatly aided in the mapping of wetland vegetation cover and distribution. Researchers would examine the photographs in order to classify the many kinds of plant life found in the marsh. Understanding the make-up of wetland ecosystems, the presence of biodiversity, and the feasibility of habitats for various species required this information.

Vegetation Mapping: Aerial photography may be used to successfully create precise vegetation maps of wetland regions. Using remote sensing techniques like supervised or unsupervised classification, we may classify the different types of vegetation that are found inside wetlands and create accurate maps that depict where this vegetation is located.

Vegetation Density and Biomass Estimation: Aerial photographs can be used to determine the biomass and plant density in a wetland. By analysing the reflectance qualities and structure of vegetation from aerial photographs, researchers may create models to predict the density and biomass of vegetation. Wetland management and conservation are greatly aided by the data generated by these models.

Vegetation Change Detection: By comparing aerial photographs taken at different dates over wetland areas, changes in the vegetation can be observed. Analysis of differences in the cover of vegetation, the species composition, or indications of the health of the vegetation allows researchers to monitor the dynamics of wetland areas and analyse the consequences of natural or human-induced changes like shifts in land use or climate change.

Invasive Species Monitoring: Aerial photography can be used to detect and track invasive plant species in wetland areas. By analysing spectral signatures and spatial patterns, researchers may detect the presence and abundance of invasive species, monitor the spread of these plants, and assess the impact they have on the native wetland vegetation (Magee 2013; Pennock et al. 2007).

4.2.1.4 Habitat Assessment

Aerial photographs may be helpful in assessing the condition of wetland habitat and determining which plant species will thrive there. The structure, patchiness, and connection of the vegetation allow researchers to assess the viability of wetland habitats for certain plant communities, endangered species, or habitat restoration initiatives.

Habitat Suitability Mapping: To determine whether or not a certain area is suited for the existence of a given kind of animal, aerial photographs can be used to create a detailed map of the area in question. By analysing habitat characteristics including the type of vegetation, the architecture of the habitat, the availability of water, and the composition of the land cover, researchers can find and map locations that have high or low suitability for a range of animal species.

Habitat Fragmentation and Connectivity: Aerial photography can help with assessments of wildlife habitat connectivity and fragmentation. By analysing the spatial layout of habitat patches and the existence of landscape features like highways, rivers, or obstacles, researchers can assess the level of habitat fragmentation and crucial corridors or connections that allow for wildlife movement and gene flow.

Nesting and Breeding Site Identification: Wildlife breeding and nesting locations can be located with the help of aerial images. Researchers can find and evaluate the quality of crucial reproductive habitats by scouring aerial pictures for characteristics like nests, burrows, and breeding ponds.

Wildlife Population Monitoring: Wildlife population estimates can be made with the help of aerial images. Researchers can keep tabs on population trends, preferred habitats, and shifts in wildlife distribution by flying overhead and counting and evaluating the number of animals.

Wildlife Behaviour and Habitat Use: Wildlife habits and habitat utilisation can be better understood through aerial photography. Researchers can learn more about how wildlife species exploit various habitat types by analysing aerial photos of animal activity, feeding patterns, or roosting sites. Wildlife habits and habitat utilisation can be better understood through aerial photography. Researchers can learn more about how wildlife species exploit various habitat types by analysing aerial photos of animal activity, feeding patterns, or roosting sites.

Invasive Species Detection and Monitoring: Invasive species in animal habitats can be detected and monitored with the help of aerial photography. Researchers can detect regions invaded by non-native species and evaluate their impact on native wildlife habitat by studying changes in vegetation composition, habitat structure, or land cover patterns.

Habitat Change Detection: By comparing aerial images taken at various times, changes in wildlife habitats can be identified. The impacts of human activities and natural processes on the dynamics of animal habitats can be understood by analysing changes in land cover, vegetation composition, and habitat features.

4.2.1.5 Land Cover Classification

Wetland classes and other land cover categories such as upland forests, agricultural fields, and urban areas were categorised by interpreters in the study region. This categorisation relied on human inspection of aerial pictures and might be improved with the use of other data sources like field surveys or preexisting land cover maps.

Open Water: Lakes, ponds, rivers, streams, and other bodies of water that remain wet year-round are included in this category of wetlands. Unvegetated patches are what this term usually describes.

Emergent Vegetation: Examples of this type of wetland include marshes, swamps, and wet meadows, all of which are dominated by herbaceous vegetation that grows above the water's surface. Cattails, reeds, sedges, grasses, and bulrushes are all examples of emergent vegetation.

Shrubs and Scrub: Shrubby wetland communities, including those dominated by willows and alders, are included in this category. These zones are frequently found in or near wetland complexes.

Forested Wetlands: This class includes wetland areas dominated by trees, such as swamp forests or bottomland hardwood forests. Common tree species found in forested wetlands include cypress, mangroves, black gum, red maple, or bald cypress. The presence of shrubs and woody vegetation, such as willows and alders, defines this category of wetland environments. Typically, you can find these zones near or inside a marsh.

Agriculture or Cropland: Some marsh areas may be used for farming or have been transformed into agriculture. Wetland rice fields, cranberry bogs, and other types of wetland-based managed agriculture fall under this category.

Urban or Built-up Areas: Sometimes, urban or built-up areas have expanded into previously natural wetland areas. Wetland ecosystems can be threatened by the expansion of human settlements, commercial and industrial infrastructure, roadways, and other locations.

Bare or Exposed Soil: Those parts of a wetland that don't have a thick layer of vegetation cover belong to this category. It's possible to see this during dry spells or in locations prone to erosion and silt deposition.

Transitional or Ecotone Areas: The ecotones found in wetlands are areas where the wetland and upland habitats meet and transition into one another. Land cover in these regions may be heterogeneous, resembling a mosaic of wetland and upland elements (Meire 2008; Olson 2012).

4.2.2 Steps of Aerial Image Assessment

Aerial photos were obtained at different intervals throughout the study, allowing researchers to detect and follow changes in wetland ecosystems. By comparing photos taken at various times over the course of several years, researchers were able to assess how wetland boundaries, vegetation cover, and land use had changed. We were able to use this data to determine if degradation, loss, or changes in wetland quality were the result of natural or anthropogenic causes. In prior decades, comparing aerial photographs from different time periods was necessary to discern changes in wetland area, vegetation cover, and land use. This allows for the detection and evaluation of wetland area shifts. The steps involved are outlined below.

Collection of Aerial Photographs: Several aerial images were collected, spanning the time frame of interest. Aeroplanes, helicopters, and other flying camera platforms are all possibilities here.

Georeferencing and Image Registration: To georeference the aerial photos, we aligned them with precise spatial reference systems or base maps. Using geolocation, the photos may be properly overlaid and compared for a reliable change detection study.

Visual Interpretation and Photointerpretation: Skilled interpreters visually analysed the aerial photographs, comparing the land cover and wetland features visible in each photograph. They looked for changes in wetland boundaries, water bodies, vegetation patterns, and other distinguishing characteristics. Georeferencing the aerial pictures meant lining them up with precise spatial reference systems or base maps. The photos could be properly overlaid and compared thanks to georeferencing, guaranteeing reliable change detection analysis.

Delineation of Wetland Boundaries: Photographs showing the extent of wetland areas at various times were annotated by interpreters to show where they began and ended. Photos might have their boundaries drawn or digitised on the spot, or the data could be exported to a computer for processing.

Change Detection Analysis: Wetland alterations were detected by comparing wetland boundaries from various time periods. Wetland loss, gain, or change could be evaluated by superimposing the wetland boundaries from the historical and contemporary images.

Quantification and Documentation: Changes in wetland extent were quantified and documented, indicating the magnitude and spatial distribution of wetland changes over time. This information was often presented as maps, statistics, or narrative reports to communicate the findings of the wetland change detection analysis. To see if there were any shifts in wetland size, we evaluated wetland boundary data from various time periods. The interpreters were able to determine which places had wetland loss, gain, or alteration by superimposing the wetland boundaries from the historical and contemporary pictures National Research Council (1993).

References

Bowman S (1990) Radiocarbon dating, vol 1. Univ of California Press

DeLaune RD, Reddy KR, Richardson CJ, Megonigal JP (2020) Methods in biogeochemistry of wetlands, vol 28. John Wiley & Sons

Duling K (2018) Carbon dating. Cavendish Square Publishing, LLC
Essington ME (2015) Soil and water chemistry: an integrative approach. CRC press
Kampa E, Hansen W (2004) Heavily modified water bodies: synthesis of 34 case studies in Europe with 19 figures and 49 tables. Springer Science & Business Media
Li D, Liu S (2018) Water quality monitoring and management: basis, technology and case studies. Academic Press
Magee T (2013) A field guide to community-based adaptation. Routledge
Meire P (ed) (2008) Integrated water management: practical experiences and case studies, vol 80. Springer Science & Business Media
National Research Council (1993) Soil and water quality: an agenda for agriculture. National Academies Press
Olson GW (2012) Field guide to soils and the environment applications of soil surveys. Springer Science & Business Media
Pennock D, Yates T, Braidek J (2007) Soil sampling designs. *Soil sampling and methods of analysis*, 1–14
Schinner F, Öhlinger R, Kandeler E, Margesin R (eds) (2012) Methods in soil biology. Springer Science & business media

Present Technologies for Wetland Studies

5

Abstract

This chapter delves into the realm of remote sensing technologies, each section unveiling a distinct methodological approach and its applications in environmental studies. The first section, "Optical Remote Sensing", elucidates the utilisation of the electromagnetic spectrum in the visible and near-infrared ranges to capture high-resolution, multispectral images of the Earth's surface. This technology proves invaluable for detailed analysis of land cover, vegetation health, and monitoring environmental changes essential for land use planning and disaster management. Moving forward, "Microwave Remote Sensing" takes centre stage, exploiting microwave frequencies to penetrate cloud cover and measure crucial environmental parameters like soil moisture and sea surface conditions. This section underscores its significance in weather forecasting, soil moisture mapping, and ocean current tracking, enhancing water resource management and climate research. The exploration continues with "Hyperspectral Remote Sensing", a technique capturing data in numerous narrow spectral bands, offering detailed insights into the Earth's surfaces' spectral signatures and material characteristics. Its applications range from mineral exploration to crop health assessment and water pollution detection. The fourth section highlights "LiDAR Technology", leveraging laser beams to create detailed, three-dimensional representations of landscapes, crucial for terrain mapping, forest monitoring, and habitat assessment for wildlife conservation. Lastly, "UAV Technology" explores the versatile use of unmanned aerial vehicles (UAVs) or drones in wetland research and conservation efforts, offering efficient aerial imagery capture and data collection for wildlife monitoring, wetland mapping, and disaster response. This unit unveils a rich tapestry of remote sensing technologies, each playing a vital role in advancing our understanding and conservation of wetland ecosystems.

L. K. Sharma, R. Naik, *Conservation of Saline Wetland Ecosystems*,
https://doi.org/10.1007/978-981-97-5069-6_5

Keywords

Hyperspectral remote sensing · LiDAR technology · Microwave remote sensing · Optical remote sensing · Unmanned Aerial Vehicles (UAV) technology

5.1 Optical Remote Sensing

5.1.1 Elements of Remote Sensing

Radiation that originates from or is reflected by things on the Earth's surface is the energy source in this context. Energy sources can be either naturally occurring, like sunlight, or artificial, such as in the case of active sensors that produce their power (e.g., RADAR).

Radiation and Interaction with the Atmosphere: The energy from the source will interact with the atmosphere once it reaches the surface of the Earth. Surface objects absorb some of the energy, some of it is scattered or reflected back into space, and the rest keeps interacting with them.

Energy Detection and Recording: Energy reflected or radiated by the Earth's surface can be detected by sensors onboard satellites, aircraft, or drones. These detectors collect information on the strength of radiation throughout a range of wavelengths, or spectral bands.

Platforms and Sensors: Sensors used to collect remote sensing data are typically mounted on platforms like satellites, aeroplanes, or drones. Optical, thermal infrared, microwave, and lidar sensors are just a few examples of the many types of sensors used to collect data across the electromagnetic spectrum.

Data Acquisition: Sensing devices collect information for remote sensing by analysing the energy reflected or emitted by the Earth's surface. The spatial and spectral information of Earth's features are preserved in digital photographs or other data forms.

Preprocessing and Calibration: The data must first be preprocessed and calibrated before it can be studied. Data correctness and consistency can be guaranteed by performing preprocessing steps such as adjusting for atmospheric influences, sensor artefacts, geometric distortions, and radiometric calibration.

Data Analysis and Interpretation: Information about the Earth's surface is gleaned through analysis and interpretation of collected and preprocessed remote sensing data. Methods from the field of image processing, such as picture editing, categorisation, detection of changes, and analysis of spatial relationships, are required.

Image Display and Visualisation: To further aid in the interpretation and transmission of the results, the data is frequently displayed as graphics or maps. The visual representation of the data is improved by the use of several techniques, such as false-colour composites, picture fusion, and 3D visualisation.

Information Application: Environmental monitoring, land cover classification, urban planning, agriculture, disaster management, and resource assessment are just few of the many uses for the interpreted remote sensing data. Decision-makers and those interested in surface-level changes can both benefit from the data gleaned by remote sensing (Arst and Arst 2003).

5.1.2 Types of Remote Sensing Sensors

In remote sensing, various sensors are used to detect and quantify electromagnetic radiation across a wide range of wavelengths. The wavelengths picked up and the level of detail provided by these detectors are two distinguishing characteristics. Some typical examples of optical sensors are shown below.

5.1.2.1 Multispectral Sensors

Multispectral sensors collect data from a wide range of the electromagnetic spectrum, including the visible, near-infrared, and even shortwave infrared areas. They supply information in a few to many spectral bands, which can be used for environmental monitoring and analysis of vegetation health and land cover. Landsat, Sentinel-2, and MODIS satellite sensors are only a few good examples.

5.1.2.2 Hyperspectral Sensors

Hyperspectral sensors are able to collect electromagnetic energy across a broad range of wavelengths by separating it into several tiny spectral bands. They offer comprehensive spectrum data, which improves the ability to identify and characterise elements on the planet's surface. Mineral exploration, crop health evaluation, and pollution monitoring are just a few of the many uses for hyperspectral data.

5.1.2.3 Panchromatic Sensors

Panchromatic sensors are able to detect light across a wide spectrum, from the visible to the near-infrared. High-resolution grayscale photography is provided by these satellites, along with precise geographical information and some spectrum discrimination. When a great deal of spatial detail is needed in a map, panchromatic data is frequently used.

5.1.2.4 Thermal Infrared Sensors

Thermal infrared sensors are able to pick up electromagnetic radiation in the infrared band that is emitted by the Earth's surface. They can be used for monitoring heat signatures, spotting thermal abnormalities, and analysing changes in land surface temperature. Common uses for these sensors include fire detection, investigation of urban heat islands, and agricultural monitoring.

5.1.2.5 LiDAR Sensors

Light Detection and Ranging (LiDAR) sensors determine the distance to the ground by sending out laser pulses, yielding precise data in three dimensions. Digital Elevation Models (DEM), topographic maps, descriptions of vegetation structures, and evaluations of landscape characteristics can all benefit from LiDAR data collection and analysis. In addition to its active sensor capabilities, LiDAR can also use an optical sensor to capture colour imagery.

Some applications of optical sensors are listed above. Depending on the parameters of the intended use, different types of sensors will have different advantages and benefits (Avtar and Watanabe 2020).

5.1.3 Optical Remote Sensing

Optical remote sensing is a powerful tool for studying wetland ecosystems and the characteristics they harbour. Sensors and other equipment that can capture and analyse electromagnetic radiation in the visible, near-infrared, and shortwave infrared ranges are used for this purpose.

Sensors that can capture and detect electromagnetic radiation in the visible, near-infrared, and shortwave infrared areas of the electromagnetic spectrum are used in optical remote sensing to collect data about the Earth's surface. Data is gathered and analysed by placing optical sensors on satellites, aeroplanes, or drones in order to see the Earth from above.

In optical remote sensing, sensors are used to detect and record the intensity and spectrum properties of light reflected or emitted from the Earth's surface. The interplay between the Sun's rays and the materials that make up the Earth's surface results in distinctive reflectance patterns across the electromagnetic spectrum (such as flora, water bodies, soil, and constructions constructed by humans, for example). The characteristics, composition, and distribution of land cover, flora, water bodies, and other things on the Earth's surface can all be gleaned from these patterns.

In digital photography used for optical remote sensing, each pixel represents a discrete point on the Earth's surface and stores spectral information. After capturing these photos, computational analysis can begin. These images can be analysed and processed to reveal details about the environment, such as land cover, plant health, urban expansion, and the effects of natural disasters.

One of the numerous benefits of optical remote sensing is that it can provide information that is synoptic, repeatable, and multispectral over large areas. It allows for the tracking of evolving processes through time, such as the development of land cover shifts, the growth of flora, and land use activities. Optical remote sensing data is used extensively in many fields, such as agriculture, forestry, environmental monitoring, urban planning, disaster management, and natural resource assessment. One of optical remote sensing's numerous benefits is the data it can provide across large areas that is synoptic, repeatable, and multispectral. It allows for the tracking of evolving processes throughout time, like the development of land cover shifts, the growth of flora, and land use activities. Agricultural, forest, environmental, urban, and disaster management, as well as natural resource assessment, all make substantial use of data gathered by optical remote sensing.

Satellites like Landsat, Sentinel, and MODIS are just a few examples of where one may find useful data for optical remote sensing. These satellites can capture imagery with a spatial resolution that ranges from the metres to the sub-meter range, allowing for multi-scale analysis. In addition, airborne platforms and Unmanned Aerial Vehicles (UAVs) equipped with optical sensors provide a higher spatial resolution and can be employed for more niche or specific purposes (Beard and McLain 2012; Chang 2007; Dong and Chen 2017).

5.1.3.1 Types of Optical Satellites

Landsat series: The Landsat satellites, operated by NASA and the United States Geological Survey (USGS), have been providing Earth observation data since 1972. The latest satellite in the series is Landsat 9, launched in September 2021.

Sentinel-2: Sentinel-2 is a mission of the European Space Agency (ESA) as part of the Copernicus program. It consists of two identical satellites, Sentinel-2A and Sentinel-2B, launched in 2015 and 2017, respectively.

SPOT series: SPOT (Satellite Pour l'Observation de la Terre) is a series of Earth observation satellites operated by the French Space Agency (CNES). The latest satellite in the series is SPOT 7, launched in 2014.

WorldView series: The WorldView satellites, operated by DigitalGlobe, provide high-resolution imagery for various applications. WorldView-1, WorldView-2, WorldView-3, and WorldView-4 are some of the satellites in this series.

Pleiades: The Pleiades satellites, operated by the CNES, are capable of capturing very high-resolution imagery. Pleiades 1A and Pleiades 1B were launched in 2011 and 2012, respectively.

GeoEye series: GeoEye-1, launched in 2008, is a high-resolution satellite operated by GeoEye (now part of DigitalGlobe). It provides imagery for mapping, monitoring, and other applications.

RapidEye: The RapidEye constellation consists of five small satellites providing multispectral imagery. These satellites were launched in 2008.

Gaofen series: The Gaofen satellites are part of China's high-resolution Earth observation program. The series includes various satellites like Gaofen-1, Gaofen-2, Gaofen-3, etc.

Cartosat series: The Cartosat satellites are operated by the Indian Space Research Organisation (ISRO). They provide high-resolution imagery for mapping and other applications. Cartosat-2, Cartosat-2A, Cartosat-2B, and Cartosat-2C are some of the satellites in this series.

KOMPSAT series: The KOMPSAT (Korean Multi-Purpose Satellite) satellites are operated by the Korea Aerospace Research Institute (KARI). This series includes satellites like KOMPSAT-2, KOMPSAT-3, KOMPSAT-3A, and KOMPSAT-5, which offer various imaging capabilities.

Examples of Optical Satellites for Wetland Studies

Many different types of wetland research have benefited from the use of these satellites, including mapping wetland extent and changes, monitoring vegetation dynamics, evaluating water quality, locating wetland habitats, and investigating wetland ecosystem processes explained diagrammatically in Fig. 5.1 and some examples in

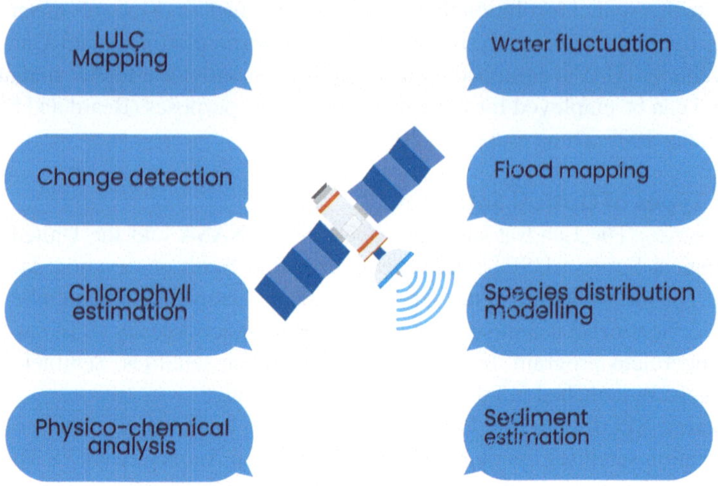

Fig. 5.1 Use of optical remote sensing for wetlands

Table 5.1. Twenty optical satellites have been deployed to study wetlands, as listed below.

Landsat series (Landsat 5, 7, 8, and 9), Sentinel-2A and Sentinel-2B, SPOT 6 and SPOT 7, WorldView-1, WorldView-2, WorldView-3, and WorldView-4, Pleiades 1A and Pleiades 1B, GeoEye-1, RapidEye constellation (RapidEye-1, RapidEye-2, RapidEye-3, RapidEye-4, and RapidEye-5), Gaofen-1, Gaofen-2, Gaofen-3, and other Gaofen satellites, Cartosat-2 series (Cartosat-2, Cartosat-2A, Cartosat-2B, and Cartosat-2C), KOMPSAT-2, KOMPSAT-3, KOMPSAT-3A, and KOMPSAT-5, COSMO-SkyMed (Constellation of Small Satellites for Mediterranean basin observation) series, ALOS (Advanced Land Observing Satellite) series (ALOS, ALOS-2), Deimos-1 and Deimos-2, SuperView-1 (also known as Gaojing-1), TripleSat constellation (TripleSat-1, TripleSat-2, and TripleSat-3), SPOT 5, IKONOS, FORMOSAT-2, ZY-3 (Ziyuan-3), DubaiSat-1 and DubaiSat-2. In Table 5.2, some examples of uses of optical satellites for saline wetlands are given.

5.1.3.2 Wetland Classification, Mapping, and Change Detection

Many different types of wetland research have benefited from the use of these satellites, including mapping wetland extent and changes, monitoring vegetation dynamics, evaluating water quality, locating wetland habitats, and investigating wetland ecosystem processes. Twenty optical satellites have been deployed to study wetlands, as listed below.

Sensor Selection: Pick a detector that works well in the spectrum you want to analyse. Multispectral or hyperspectral sensors are commonly used in wetland mapping projects. Airborne sensors like hyperspectral imaging spectrometers and satellite-borne sensors like Landsat, Sentinel-2, and MODIS are a few examples.

Table 5.1 Examples of optical satellites used for wetland study

S. no.	Name of wetland	Type of wetland	Optical satellite used	Purpose of use
1	Everglades	Marshland	Landsat 8	Monitoring wetland changes
2	Okavango Delta	Inland Delta	Sentinel-2	Assessing vegetation and water extent
3	Pantanal	Seasonal Floodplain	WorldView-3	Biodiversity monitoring
4	Sundarbans	Mangrove Forest	QuickBird	Coastal change detection
5	Danube Delta	River Delta	Copernicus Sentinel-2	Monitoring wetland ecosystems
6	Kakadu National Park	Wetland	Landsat 8	Vegetation mapping
7	Hula Valley	Marshland	Landsat 7	Wetland health assessment
8	Tonle Sap Lake	Flooded Forest	WorldView-2	Water quality and habitat monitoring
9	Lake Chad	Endorheic Lake	Sentinel-2	Water level and land use analysis
10	Okefenokee Swamp	Swamp	Landsat 5	Wetland habitat changes
11	Liaohe Delta	River Delta	Gaofen-1	Coastal wetland monitoring
12	Narew National Park	Wetland	Landsat 7	Ecosystem health assessment
13	Tisza River Basin	River Basin	Sentinel-2	Monitoring water quality
14	Morava River Floodplain	Floodplain	Pleiades	Flood risk assessment
15	Esteros del Ibera	Wetland	Landsat 8	Biodiversity and land use analysis
16	Poyang Lake	Seasonal Lake	Sentinel-2	Water level and habitat monitoring
17	Rann of Kutch	Seasonal Salt Marsh	WorldView-3	Coastal and wetland changes
18	Itaipu Dam Reservoir	Reservoir	Landsat 5	Reservoir water level monitoring
19	Humber Estuary	Estuary	Copernicus Sentinel-2	Coastal zone and water quality analysis

Preprocessing: Prepare the remotely sensed data by adjusting for sensor-specific artefacts and removing atmospheric effects. To provide accurate and comparable data, standard preprocessing stages may include atmospheric adjustment, radiometric calibration, and geometric correction.

Image Interpretation: Find the wetland areas in the final images by using your keen eyesight. Wetlands have their own distinct spectral fingerprints that distinguish them from other types of land cover. Wetland characteristics, such as vegetation types, water bodies, and specific spectral responses, are taken into account while interpreting images. Examine the finalised images with a keen eye for the wet-

Table 5.2 Use of optical satellites for saline wetland

S. no.	Name of saline wetland	Type of saline wetland	Optical satellite used	Purpose of use
1	Chott el Jerid	Salt Flat	Landsat 8	Monitoring salt flat dynamics
2	Sambhar Lake	Salt Lake	WorldView-3	Mapping lake surface and salt deposits
3	Great Salt Lake	Terminal Lake	Sentinel-2	Assessing lake level and water dynamics
4	Lake Urmia	Endorheic Lake	PlanetScope	Monitoring lake area and salt content
5	Maranjab Salt Desert	Salt Desert	Sentinel-2	Mapping salt desert features
6	Etosha Pan	Salt Pan	WorldView-3	Studying pan conditions and wildlife
7	Lake Natron	Alkaline Lake	Landsat 8	Mapping lake and shoreline changes
8	Salar de Uyuni	Salt Flat	Sentinel-2	Assessing salt flat topography and changes
9	Lake Eyre	Salt Pan	WorldView-3	Topographic mapping and flood assessment
10	Lop Nur	Salt Pan	PlanetScope	Monitoring salt pan conditions and topography
11	Badwater Basin	Salt Pan	Landsat 8	Assessing salt pan topography and changes
12	Lake Abert	Endorheic Lake	WorldView-3	Monitoring lake water quality and sediments
13	Lake Tuz	Salt Lake	Sentinel-2	Mapping lake chemistry and conditions
14	Koyashskoye Salt Lake	Salt Lake	PlanetScope	Assessing salt lake chemistry and minerals
15	Sebkha Sejoumi	Salt Pan	WorldView-3	Studying salt pan features and composition
16	Lake Tengiz	Salt Lake	Sentinel-2	Monitoring lake water chemistry and sediments
17	Lake Qarun	Salt Lake	PlanetScope	Mapping lake conditions and shoreline
18	Kara-Bogaz-Gol	Lagoon	WorldView-3	Assessing lagoon water quality and features
19	Soda Lakes of Kenya	Saline Lakes	Sentinel-2	Studying saline lake chemistry and sediments
20	Lake Assal	Crater Lake	Landsat 8	Mapping crater lake chemistry and sediments

land habitats. Wetlands have their own distinct spectral fingerprints that distinguish them from other types of land cover. Wetland characteristics, such as vegetation types, water bodies, and specific spectral responses, are taken into account while interpreting images.

Classification Algorithms: Wetlands can be automatically distinguished from other land cover categories by employing classification algorithms. It's not uncommon to employ either supervised or unsupervised classification strategies. Unlike unsupervised classification, which classifies pixels based on their spectral similarity without prior knowledge, supervised classification trains the algorithm with a set of representative training samples.

Change Detection Algorithms: Use algorithmic change detection to locate and count deviations from the registered images. Object-based methods, which take changes at the object or patch level into account, complement pixel-based methods like picture differencing and change vector analysis.

Change Detection Thresholding: To separate out meaningful changes from background noise and random fluctuations, thresholds and other change detection criteria must be defined. A combination of statistical analysis and expert knowledge is often used to arrive at threshold levels. Normalised Difference Vegetation Index (NDVI) and Normalised Difference Water Index (NDWI) are two examples of change detection indices that can be used to zero in on specific types of shifts.

Change Analysis and Interpretation: Examine the observed shifts in wetland features and provide an explanation for them. This entails analysing land use, hydrological changes, and climatic change in relation to identified changes, as well as determining the extent and significance of such changes. The discovered alterations can be better understood with the use of expert knowledge and ground truth data. Examine the observed shifts in wetland features and explain them. This entails analysing land use, hydrological changes, and climatic change in relation to identified changes, as well as determining the extent and significance of such changes. The discovered alterations can be better understood with the use of expert knowledge and ground truth data.

Validation: Verify the precision of the wetland classification by contrasting it to independent sources of information. Reference data for validation might be gathered by field surveys or high-resolution imaging. The trustworthiness of the classification findings is assessed using accuracy assessment methods like error matrices and kappa statistics.

Post-Classification Processing: Improve the quality and readability of the wetland maps by doing post-classification processing, such as filtering, smoothing, and aggregation, on the classification findings. Digital elevation models and other supplementary data can be included to improve categorisation accuracy (Elachi and Van Zyl 2021; Fahlstrom et al. 2022; Liang and Wang 2019).

5.1.3.3 Wetland Mapping

Make wetland maps using the data from the classification. Raster pictures, in which each pixel represents a wetland class, and vector polygons, which outline wetland boundaries, are also acceptable formats for these maps. Information on wetland distribution, fragmentation, and size can be gleaned from wetland maps through additional analysis.

5.1.3.4 Vegetation Analysis

The three most common types of plant life in wetland ecosystems are emergent vegetation, submerged vegetation, and floating vegetation. Optical remote sensing can be used to determine important aspects of wetland ecosystems like vegetation cover, species composition, health, and biomass. Vegetation indices, such as the Normalised Difference Vegetation Index (NDVI), can be calculated using remotely sensed data and used to assess the health and resilience of plant life. Optical remote sensing, using satellites or aerial sensors, enables vegetation study in wetland regions. This opens the door to evaluating the wetland's plant cover, species composition, health, and biomass. The steps involved in analysing vegetation are as follows:

Vegetation Indices: Vegetation indices that measure vitality, density, and stress can be calculated. The NDVI is the most used vegetation index, and it is derived by taking the inverse of the ratio of near-infrared (NIR) reflectance to red reflectance. Higher NDVI values, which range from −1 to 1, indicate more robust plant life.

Species Discrimination: Differentiate between species of wetland vegetation or plant functional kinds using spectral signatures and vegetation indicators. Vegetation in wetlands can be either floating or submerged, or it can be both. It is feasible to identify and map the spread of various types of vegetation by examining their spectral responses.

Change Detection: Compare numerous satellite or aerial photographs or photographs taken at different times to track the development of wetland vegetation through time. Wetland vegetation changes, such as those caused by invasive species or human disturbance, can be detected by measuring shifts in vegetation indices or spectral reflectance.

Biomass Estimation: Understanding the carbon storage capacity and overall ecosystem productivity of wetlands requires an accurate estimate of their vegetation biomass. Vegetation indices, such as the Enhanced Vegetation Index (EVI) and the Normalised Difference Moisture Index, can be derived from biomass using optical remote sensing (NDMI). Wetland biomass can then be estimated by calibrating and validating these indices against ground-based data.

Wetland Health Assessment: Wetland vegetation health and stress can be evaluated with optical remote sensing data. Vegetation stress caused by things like waterlogging, nutrient inadequacy, or pollution can be detected through changes in vegetation indices over time, such as a decline in NDVI. Vegetation indicators' temporal dynamics can be analysed to pinpoint problem regions in wetland ecosystems.

5.1.3.5 Water Quality Monitoring

Wetland water quality metrics like turbidity, chlorophyll-a concentration, dissolved organic matter, and suspended sediments can be monitored via optical remote sensing. Water quality indicators can be derived from remotely sensed data, allowing for the detection of pollution sources, algal blooms, and temporal shifts in water chemistry.

Optical remote sensing is used to monitor water quality in lakes, rivers, and other water bodies by assessing a variety of water quality indicators using satellite or airborne sensors. Clarity, chlorophyll-a content, and suspended particles are only few of the markers of water quality that can be gleaned using optical remote sensing. Here are the main steps of optical remote sensing for monitoring water quality:

Water Clarity Assessment: Determine the water body's clarity and transparency by analysing its spectral reflectance qualities. Turbidity, a measure of the amount of suspended particles in the water, can be calculated by analysing the correlation between reflectance in different spectral bands, such as blue and green. High turbidity levels indicate cloudy water.

Chlorophyll-a Concentration: Use optical remote sensing data to calculate chlorophyll-a concentration, a measure of phytoplankton population and water quality. The green portion of the spectrum is reflected by chlorophyll-a while the blue and red portions are absorbed. These spectral features can be used by algorithms applied to satellite or airborne data to infer chlorophyll-a concentrations.

Suspended Sediment Concentration: Determine the amount of suspended silt in water by measuring its reflectance. When there is a lot of silt present, the light is scattered and absorbed in diverse ways, producing a unique spectrum of reflectance. These trends can be utilised as inputs for algorithms that predict suspended sediment concentrations.

Harmful Algal Blooms (HABs): Optical remote sensing data can be used to identify and track toxic algal blooms. Toxins produced by some algae species can be hazardous to the health of humans and other aquatic animals. The presence of algal pigments in the water can be detected by optical sensors. HABs are detected and tracked using specialised algorithms and spectral bands.

Water Temperature: Check the water temperature via infrared thermography. Satellites equipped with thermal sensors can determine water temperature by measuring the thermal radiation emitted by bodies of water. Understanding thermal gradients and temperature fluctuations in water bodies is important for understanding the health of aquatic ecosystems, and this data can help.

5.1.3.6 Hydrological Studies

Hydrological stability and water flow regulation are dependent on wetland ecosystems. Wetland hydrology can be evaluated with the help of remote sensing by keeping an eye on things like surface water extent, water level variations, and the dynamics of floods. Wetland hydrological processes and their reaction to climate change and human activities can benefit from the knowledge provided by this research.

The hydrological equilibrium of the planet depends on the function of wetlands. By keeping an eye on surface water extent, water level variations, and flood dynamics, remote sensing can help with wetlands hydrology assessments. The reaction of wetland hydrological processes to climate change or human activities can be better understood with the help of this data.

Surface Water Mapping: Wetlands can be mapped for their size and surface water distribution using optical remote sensing data. Water bodies can be detected with regularity using moderately-resolution satellite sensors like Landsat or Sentinel-2. Open water can be distinguished from other land cover categories by evaluating its spectral properties.

Water Level Fluctuations: Observe changes in wetland water levels using information gathered by optical remote sensing. Water level fluctuations can be seen over time by comparing aerial or satellite images taken at different times. Wetland hydrological dynamics, such as seasonal changes or long-term trends, can be better understood by the temporal analysis of water borders or inundation extents.

Flood Mapping: Optical remote sensing is a powerful tool for mapping wetland areas and monitoring floods. When floods occur, they can cause significant changes in the inundation patterns and water extent experienced by wetland areas. These changes can be detected by optical sensors by comparing before- and after-flood photos. Mapping floods can give useful data for evaluating flood risk, planning emergency actions, and administering wetland regions.

Wetland Connectivity: Analyse the hydrological links within wetland systems using optical remote sensing data. Various wetland ecosystems are often linked together by surface water flows and hydrological channels. Investigating water flow, stream networks, and drainage patterns can help us comprehend the interconnectedness of wetland areas and their interaction with the water bodies in their immediate surroundings. Analyse the hydrological links within wetland systems using optical remote sensing data. Wetland ecosystems are often interconnected by hydrological channels and surface water flows. Water flow, stream networks, and drainage patterns can be studied to get insight into the interconnectedness of wetland areas and their interaction with the water bodies in their immediate surroundings.

Water Storage and Water Balance: Wetland hydrology can only be understood in its whole through the estimation of water storage and the analysis of the water balance's constituent parts. Optical remote sensing data can help with the estimation of water volume and water balance parameters like evapotranspiration and precipitation. By integrating data from remotely sensed sensors with models of hydrological processes, it is possible to gain insight into the water budget of wetlands.

Wetland Change Detection: Watch for changes in the hydrological parameters of wetland areas using data from optical remote sensing. By comparing previous photographs or time series data, wetland decline, expansion, or changes in hydrological conditions can be detected. Examining the variations helps with this. Understanding the impacts of human activities, climate change, and shifting land use on wetland hydrology can benefit greatly from the data provided here (Marghany 2016; Megie 1985; Miller et al. 2005).

5.1.3.7 Biodiversity Assessment

Numerous plant and animal species rely on wetland habitats for survival. Wetland biodiversity may be evaluated with the help of remote sensing since it can map habitats, identify essential biological features, and track the distribution and abundance of species. Optical sensors can provide spatial and spectral data that can be used to infer the existence of particular ecosystems or indicator species.

Optical remote sensing is used for biodiversity evaluation in wetlands by analysing variables like habitat diversity, species richness, and vegetation shape through the use of satellite or airborne sensors. Optical remote sensing may not be able to evaluate species composition directly, but it can reveal important details about the wetland's physical features that are linked to biodiversity. The following are the main steps in optical remote sensing for wetland biodiversity assessment:

Habitat Mapping: Using optical remote sensing data, draw a detailed map of the marsh's varying ecosystems. Studying the spectral features and spatial patterns of vegetation allows for the differentiation of different forms of land cover, such as forests, marshes, open water, and bare soil. The location and composition of wetland biodiversity can be better understood with the help of a habitat map.

Structural Analysis: Use information gathered via optical remote sensing to examine plant structure. Using remote sensing, we may estimate parameters like vegetation height, canopy cover, and Leaf Area Index (LAI). The complexity of a habitat, which is in turn related to biodiversity, can be gleaned from its structural characteristics. Wetlands, for instance, may be able to house more species if they have a varied vegetation structure.

Landscape Metrics: Using landscape metrics, you can evaluate the wetland's spatial layout and configuration of habitat patches. Patch size, form index, and edge density are all landscape metrics used to measure the degree to which ecosystems are fragmented or connected. The impact of habitat heterogeneity on biodiversity can be measured with these indicators.

Species Distribution Modelling (SDM): Model the distribution of target indicator species using information gleaned through optical remote sensing in addition to other factors including climate, geography, and hydrological characteristics. Species distribution models use environmental variables to make predictions about the occurrence and abundance of species. Information on habitat suitability, vegetation structure, or landscape context can all be incorporated into these models with the help of remote sensing data.

5.1.3.8 Land Use Land Cover (LULC) Classification

Several different classification techniques can be used for LULC mapping in wetland regions utilising optical remote sensing. The study's goals, the wetland's characteristics, and the classification system all have a role in determining the exact categories employed. Types of LULC identified by optical remote sensing in wetland environments include the following:

Open Water: Lakes, ponds, rivers, and other bodies of water are all included in this category of wetland terrain. It usually describes places where there is little to no cover from newly emerged vegetation.

Emergent Vegetation: Surface herbaceous plant life is characteristic of wetlands such as marshes, swamps, and wet meadows. Wetlands like reed beds, cattail marshes, and even grasslands fall under this category.

Woody Vegetation: This type of wetland is characterised by the presence of woody vegetation, such as shrubs and trees. Wooded wetlands, such as mangrove and swamp forests, also fall into this category, as do shrubby wetlands such as willow and alder thickets.

Bare Soil or Sediment: Wetlands can have patches of exposed soil or sediment due to erosion, deposition, or ebb and flow of water. Vegetation cover is typically low in these regions, hence their spectral features may differ from those of other classes.

Agriculture or Cropland: Croplands are often created inside wetland ecosystems, suggesting that some wetlands have been converted to agricultural use. Wetland rice fields, cranberry bogs, and other forms of managed agriculture fall under this category.

Urban or Built-up Areas: The conversion of wetland areas into urban or built-up environments is one way in which urbanisation affects wetlands. In this category, we find anything from homes and businesses to highways and factories that are encroaching into wetland regions.

Transitional Zones: Ecotones can be found in wetlands, and they mark the slow transition between marsh and upland environments. Land cover types in these zones may be heterogeneous, signifying the transition zone between marsh and dry lands.

5.2 Microwave Remote Sensing

Remote sensing with microwaves is a strong and flexible tool that has greatly improved our ability to study the Earth's surface, atmosphere, and oceans. Data and images are gathered from spacecraft, aeroplanes, or ground-based equipment using microwave radiation, with wavelengths typically between 1 mm and 1 m. Microwave radiation is superior to visible light in that it can see through a variety of atmospheric obstacles, including clouds, rain, snow, and even vegetation. Microwave remote sensing's ability to reveal surface features and ambient conditions is a major benefit. Such details can be difficult to ascertain using conventional methods. Soil moisture, sea surface temperature, ocean currents, and polar ice sheet thickness are just a few of the things it can detect. Weather monitoring, particularly the detection of precipitation and the tracking of severe storms is another area where microwave sensors prove invaluable.

Microwave remote sensing is extremely useful in agriculture for keeping tabs on crops and estimating harvests so farmers may better manage their water and fertiliser resources. Furthermore, it helps in the identification and evaluation of natural disasters like floods and landslides. Microwave remote sensing also plays an important role in Earth science, particularly in the areas of climate study and environmental observation. Scientists use data from microwave sensors to investigate the Earth's carbon and water cycles, follow deforestation and land use changes, and comprehend the dynamics of polar regions.

In addition, microwave remote sensing is crucial for understanding Earth's polar areas since it allows researchers to track the movement of ice sheets and glaciers. It's essential for analysing the effects of climate change on polar ecosystems, as it sheds light on sea ice extent and thickness. When it comes to studying the environment, the weather, farming, and climate, microwave remote sensing has become an indispensable instrument. Its versatility in collecting data under a wide range of settings has helped us make wiser choices, more efficiently manage scarce resources, and better understand the myriad forces at work in the world. Microwave remote sensing has many potential uses, and as the field develops, we may expect to learn even more about Earth's dynamic systems (Mishra et al. 2017).

5.2.1 Use of Microwave Remote Sensing for Wetland Studies

Wetland studies can benefit greatly from microwave remote sensing because of the wealth of information it provides on wetland dynamics and features. Key applications of microwave RS in wetland research include the following which are also diagrammatically represented in Fig. 5.2 and some examples in Table 5.3 for wetlands and in Table 5.4 for saline wetlands.

Wetland Mapping and Classification Microwave remote sensing is very useful for mapping and classifying wetland areas, especially in places with dense vegetation or persistent cloud cover. Microwaves can be used to detect and map wetlands since their signals can penetrate both plants and clouds. Microwave wetland mapping is most commonly accomplished with the aid of Synthetic Aperture Radar (SAR) sensors. Satellites like Sentinel-1 and RADARSAT are equipped with such sensors.

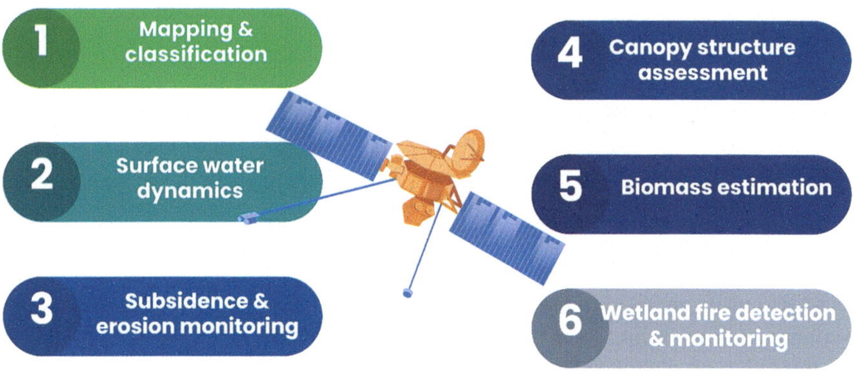

Fig. 5.2 Use of microwave remote sensing for wetland studies

Table 5.3 Use of microwave remote sensing for wetland studies

S. no.	Name of wetland	Type of wetland	Microwave remote sensing satellite	Purpose of use
1	Everglades	Marshland	Sentinel-1	Monitoring wetland water levels
2	Okavango Delta	Inland Delta	RADARSAT-2	Assessing wetland hydrodynamics
3	Pantanal	Seasonal Floodplain	ALOS-2	Monitoring flooding patterns
4	Sundarbans	Mangrove Forest	TerraSAR-X	Mapping mangrove cover changes
5	Danube Delta	River Delta	Copernicus Sentinel-1	Tracking deltaic land changes
6	Kakadu National Park	Wetland	ENVISAT ASAR	Studying wetland vegetation dynamics
7	Hula Valley	Marshland	Radarsat-2	Monitoring water levels and land use
8	Tonle Sap Lake	Flooded Forest	Sentinel-1	Tracking lake water level changes
9	Lake Chad	Endorheic Lake	ALOS PALSAR	Mapping lake dynamics and shoreline
10	Okefenokee Swamp	Swamp	TerraSAR-X	Studying wetland vegetation changes
11	Liaohe Delta	River Delta	RADARSAT-2	Monitoring deltaic land dynamics
12	Narew National Park	Wetland	Sentinel-1	Tracking wetland water levels
13	Tisza River Basin	River Basin	Advanced land observing satellite (ALOS)-2	Monitoring river basin dynamics
14	Morava River Floodplain	Floodplain	RADARSAT-2	Studying floodplain changes
15	Esteros del Ibera	Wetland	RADARSAT-2	Mapping wetland changes
16	Poyang Lake	Seasonal Lake	Sentinel-1	Monitoring lake water level changes
17	Rann of Kutch	Seasonal Salt Marsh	RADARSAT-2	Studying salt marsh dynamics
18	Itaipu Dam Reservoir	Reservoir	TerraSAR-X	Monitoring reservoir water levels
19	Humber Estuary	Estuary	RADARSAT-2	Mapping estuary water dynamics

SAR sensors are used in microwave remote sensing projects to map and classify wetland areas according to radar backscatter characteristics. Due to its capacity to see through both plants and clouds, microwave remote sensing is ideal for wetland mapping and classification.

Here are the key steps involved in wetland mapping and classification using microwave remote sensing:

Table 5.4 Examples of microwave satellite for saline wetlands

S. no.	Name of saline wetland	Type of saline wetland	Microwave satellite used	Purpose of use
1	Chott el Jerid	Salt Flat	Sentinel-1	Monitoring salt flat dynamics
2	Sambhar Lake	Salt Lake	RADARSAT-2	Mapping lake surface and salt deposits
3	Great Salt Lake	Terminal Lake	TerraSAR-X	Assessing lake level and water dynamics
4	Lake Urmia	Endorheic Lake	ALOS PALSAR	Monitoring lake area and salt content
5	Maranjab Salt Desert	Salt Desert	RADARSAT-2	Mapping salt desert features
6	Etosha Pan	Salt Pan	COSMO-SkyMed	Studying pan conditions and wildlife
7	Lake Natron	Alkaline Lake	Sentinel-1	Mapping lake and shoreline changes
8	Salar de Uyuni	Salt Flat	RADARSAT-2	Assessing salt flat topography and changes
9	Lake Eyre	Salt Pan	TerraSAR-X	Topographic mapping and flood assessment
10	Lop Nur	Salt Pan	Sentinel-1	Monitoring salt pan conditions and topography
11	Badwater Basin	Salt Pan	RADARSAT-2	Assessing salt pan topography and changes
12	Lake Abert	Endorheic Lake	ALOS PALSAR	Monitoring lake water quality and sediments
13	Lake Tuz	Salt Lake	TerraSAR-X	Mapping lake chemistry and conditions
14	Koyashskoye Salt Lake	Salt Lake	RADARSAT-2	Assessing salt lake chemistry and minerals
15	Sebkha Sejoumi	Salt Pan	COSMO-SkyMed	Studying salt pan features and composition
16	Lake Tengiz	Salt Lake	RADARSAT-2	Monitoring lake water chemistry and sediments
17	Lake Qarun	Salt Lake	TerraSAR-X	Mapping lake conditions and shoreline
18	Kara-Bogaz-Gol	Lagoon	COSMO-SkyMed	Assessing lagoon water quality and features
19	Soda Lakes of Kenya	Saline Lakes	RADARSAT-2	Studying saline lake chemistry and sediments
20	Lake Assal	Crater Lake	Sentinel-1	Mapping crater lake chemistry and sediments

Data Acquisition: Collect SAR information from spacecraft like Sentinel-1 and RADARSAT. Single (HH) or dual (HH/VV or HV) polarisation microwave imagery is available from SAR sensors. More information regarding the scattering properties of wetland surfaces can be captured by using dual-polarisation data, hence it is often favoured for wetland mapping.

Preprocessing: Improve the image quality by performing preprocessing on the SAR data to get rid of sensor-specific artefacts. To provide precise georeferencing, radiometric calibration, speckle filtering, and geometric correction are all necessary steps.

Backscatter Analysis: Use the SAR data's backscatter characteristics to locate swamps. Radar backscatter from wetlands is often easily distinguishable from that of dry ground because of the presence of plant, water, and soil moisture. Backscatter is typically strong in wetland areas with dense vegetation due to the volume scattering effect, and low in open water areas with a smooth water surface.

Polarimetric Decomposition: To learn more about the scattering processes present in wetland environments, polarimetric decomposition can be applied to the SAR data. This requires dissecting the dual-polarisation data into surface scattering, double bounce, and volume scattering components. These features aid in wetland categorisation because they reveal information about the marsh surface's physical qualities.

Classification Algorithms: Wetland regions can be identified and separated from other land cover types by applying classification algorithms to the processed SAR data. Both supervised and unsupervised classification algorithms can be used. Wetlands and non-wetlands are used to train a classifier, such as a maximum likelihood, Support Vector Machine (SVM), or random forest, in supervised classification. Wetland class labels can be automatically assigned using clustering in unsupervised classification.

Ancillary Data Integration: Accurate wetland mapping and categorisation can be achieved through the incorporation of other data sets, such as optical images, DEMs, or land cover maps. Wetland vegetation and water quality can be studied in greater depth with the aid of optical imaging, while topography characteristics and wetland hydrological connections can be better understood with the aid of DEMs.

Accuracy Assessment: Use ground truth data obtained from field surveys or high-resolution photography to verify the accuracy of the wetland mapping and categorisation results. Wetland inventories, vegetation surveys, and wetland habitat assessments are all examples of field data. The precision and accuracy of the wetland classifications are measured using accuracy assessment methods including error matrices and kappa statistics.

Wetland Mapping Products: Create wetland mapping products including wetland extent and type maps based on the findings of the classification. These maps detail the location and makeup of wetland regions throughout the research region. Wetland metrics including extent, fragmentation, and closeness to other land cover types can be derived from examined mapping products.

The inventorying, monitoring, and protection of wetland ecosystems benefit from microwave remote sensing wetland mapping and classification. Microwave sensors can see through plants and clouds, enabling precise wetland delineation even under difficult conditions (Ollero and Maza 2007; Pandey et al. 2020a, b).

Wetland Hydrology and Water Dynamics Wetland hydrology and water dynamics can be tracked with the help of microwave remote sensing. The quantity of surface water, variations in water level, and hydrological connectivity within wetland systems are all things that SAR sensors can tell us. Surface water conditions can be monitored, such as wetland flooding, drought, and other changes, by analysing the microwave signals that have been backscattered from the water.

Microwave remote sensing, in particular the processing of SAR data, can be used to effectively monitor and study wetland hydrology and water dynamics. Wetland water characteristics and dynamics can be gleaned from microwave signals since they can pass through clouds and vegetation.

Here are the key applications of microwave remote sensing for wetland hydrology and water dynamics:

Surface Water Mapping: It is possible to create a map of wetland surface water using microwave remote sensing. Because SAR data can tell the difference between water and other land cover types, wetland inundation trends can be identified and monitored. The geographical and temporal fluctuations in wetland surface water can be evaluated by evaluating the backscatter intensity and polarisation properties of SAR data.

Wetland Water Level Monitoring: Wetland water levels can be tracked over time with the help of microwave remote sensing. SAR data's backscatter intensity can be used to identify rises and falls in water levels. We can see seasonal oscillations and responses to precipitation or hydrological events in wetlands by comparing SAR images obtained at different times, allowing us to trace changes in water levels.

Wetland Water Quality Assessment: Certain characteristics of wetland water quality are observable by microwave remote sensing. Surface water roughness, for instance, is related to water quality metrics like suspended sediments or turbidity, and can be estimated using SAR data. The evaluation of wetland water conditions is made possible by monitoring changes in backscatter intensity and polarisation properties.

Wetland Inundation Monitoring: Wetland inundation dynamics, such as the duration and timing of inundation occurrences, can be tracked with the help of microwave remote sensing. Wetlands that are prone to floods or extended submergence can be pinpointed by examining how the properties of the backscatter have changed over time. This data is essential for figuring out how wetland hydrological regimes influence wetland ecosystems.

Wetland Hydrological Connectivity: Wetland hydrological connection can be better understood with the help of microwave remote sensing. The backscatter features of SAR data can be used to detect and track the waterways that connect different wetlands, rivers, and other bodies of water. This data is critical for comprehending the role of hydrological connectivity in wetland ecosystems, as well as their hydrological functioning, water exchange mechanisms, and water exchange processes.

Wetland Drying and Wetting Patterns: Wetland drying and wetting patterns can be detected and monitored via microwave remote sensing. Areas undergoing wetting or drying processes throughout time can be identified by evaluating variations in backscatter intensity and polarisation characteristics. This data is useful for figuring out how plants in wetlands react to shifting water levels and how wetlands function hydrologically.

Wetland Soil Moisture Estimation: Wetland soil moisture content estimates can be improved with the help of microwave remote sensing. Longer wavelength microwaves (L-band, C-band) can reach below the soil's surface and measure moisture levels. Microwave remote sensing of soil moisture can shed light on the dynamics of wetland water regimes, plant water availability, and water storage.

Wetland management, hydrological modelling, and conservation planning can all benefit from microwave remote sensing's ability to monitor hydrology and water dynamics across broad spatial and frequent temporal scales.

Soil Moisture Estimation Soil moisture content in wetlands can be accurately estimated by the use of microwave remote sensing. Longer wavelength microwaves (L-band, C-band) are able to seep through vegetation and reveal soil moisture status. Microwave remote sensing of soil moisture can shed light on the dynamics of wetland water regimes, plant water availability, and water storage.

Microwave remote sensing can be used to estimate soil moisture by retrieving information about the soil's surface moisture content from SAR data. Due to the microwaves' ability to travel through the atmosphere and interact with the soil surface, this kind of remote sensing is particularly useful for estimating soil moisture.

Here are the key steps involved in soil moisture estimation using microwave remote sensing:

Data Acquisition: Collect SAR information from spacecraft like Sentinel-1 and RADARSAT. Single (HH) or dual (HH/VV or HV) polarisation microwave imagery is available from SAR sensors. Soil moisture estimate often makes use of dual-polarisation data because it provides more information on the scattering qualities of the soil surface.

Preprocessing: The SAR data should be preprocessed to get rid of noise and fix sensor-specific abnormalities. Radiometric calibration, speckle filtering to mitigate speckle in images, and geometric correction for pinpoint precision in georeferencing are all part of the process.

Backscatter Analysis: The soil moisture content can be estimated by analysing the SAR data's backscatter properties. Soil dielectric characteristics, which are affected by soil moisture, have an effect on the backscatter intensity measured by the SAR sensor. Backscatter signals are enhanced by wet soil because of its greater dielectric constant.

Soil Moisture Retrieval Models: To calculate an approximation of the soil moisture from the backscatter data, use a model or algorithm designed for this purpose.

Calibration utilising ground-based observations is often used to generate correlations between backscatter and soil moisture for use in these models. The Water Cloud Model, the Dubois Model, and the Integral Equation Model are all frequently employed models.

Ancillary Data Integration: Improve the precision of soil moisture estimation by using supplementary data such as soil texture maps, vegetation indices, or meteorological information. The effect of soil type on the correlation between backscatter and soil moisture can be taken into account with the aid of soil texture data. Vegetation cover can affect the backscatter signals, and vegetation indices can reveal this. Soil moisture estimates can be put into perspective with the help of meteorological data like precipitation or evaporation.

Validation: Use in situ measurements to verify the accuracy of the soil moisture estimations calculated from the satellite data. The accuracy and trustworthiness of the remote sensing-based estimates can be determined by comparing them to ground truth data, such as in situ soil moisture measurements from soil moisture probes or sampling. Agreement between remote sensing and ground-based measurements is typically quantified using statistical analysis, such as correlation or root mean square error.

Spatial and Temporal Analysis: Examine the soil moisture distribution in terms of both space and time using satellite-based estimates. To get a sense of how soil moisture is distributed over the area under study, maps can be made. Understanding the seasonal changes, trends, and dynamics of soil moisture is important for hydrological modelling, agricultural management, and water resource planning, all of which can benefit from a temporal analysis.

Microwave remote sensing soil moisture estimation gives useful data on soil moisture conditions across broad areas and at regular intervals. It has several practical uses, including drought tracking, water supply management, flood prediction, and agricultural preparation.

Wetland Vegetation Monitoring Microwave remote sensing can provide information about wetland vegetation characteristics, such as biomass, vegetation water content, and vegetation structure. Microwave signals interact differently with vegetation depending on its density, height, and moisture content. These interactions can be analysed to estimate parameters related to vegetation health and dynamics. Estimating soil moisture via microwave remote sensing is useful because it allows for monitoring of soil moisture conditions over wide areas and at regular intervals. It can be put to use in many settings, including agricultural planning, flood forecasting, irrigation management, and drought monitoring.

SAR data is analysed to determine how microwaves interact with wetland vegetation as part of microwave remote sensing for wetland vegetation monitoring. The capacity of microwave remote sensing to see through plants and provide data on vegetation features makes it ideal for monitoring vegetation in wetlands.

Here are the key steps involved in wetland vegetation monitoring using microwave remote sensing:

Data Acquisition: Collect SAR information from spacecraft like Sentinel-1 and RADARSAT. Single (HH) or dual (HH/VV or HV) polarisation microwave imagery is available from SAR sensors. Vegetation monitoring frequently makes use of dual-polarisation data due to the data's capacity to capture more information on the scattering properties of vegetation.

Preprocessing: The SAR data should be preprocessed to get rid of noise and fix sensor-specific abnormalities. Radiometric calibration, speckle filtering to mitigate speckle in images, and geometric correction for pinpoint precision in georeferencing are all part of the process.

Backscatter Analysis: Examine the SAR data's backscatter qualities to learn more about plant life in wetland areas. Heavily vegetated wetlands have distinct dispersion properties because of the presence of leaves, stems, and branches. It is possible to estimate parameters associated with the health and dynamics of vegetation by analysing the backscatter intensity and polarisation features.

Polarimetric Decomposition: To identify the various scattering mechanisms at work in wetland vegetation, polarimetric decomposition can be applied to the SAR data. To perform polarimetric decomposition, the data from two polarisations must be separated into their individual scattering components. These factors reveal information on the biomass, water content, and structural organisation of wetland plants.

Vegetation Parameter Estimation: Use the results of the backscatter and polarimetric decomposition to make estimates of the properties of the vegetation. Vegetation metrics including biomass, water content, height, and density may all be calculated from field observations. Wetland vegetation health, structure, and dynamics can all be gleaned from these measurements.

Calibration and Validation: Use field data or ground observations to calibrate the SAR-derived vegetation parameters. Relationships between the remote sensing-derived vegetation metrics and the ground truth measurements can be established via field data like vegetation surveys, biomass sampling, or vegetation indices. Remotely sensed vegetation information can be evaluated for accuracy and reliability using validation methods including correlation analysis and root mean square error.

Temporal Analysis: Use several SAR photos taken at different times of day to study how wetland vegetation changes over time. Wetland vegetation has seasonal dynamics that can be better understood through the use of temporal analysis, such as growth, phenology, and responses to environmental changes. Wetland vegetation change can be monitored using time series analysis methods like vegetation indices and change detection algorithms.

Integration with Optical Remote Sensing: The information regarding wetland vegetation can be improved by combining microwave remote sensing data with optical remote sensing data, such as multispectral or hyperspectral photography. Vegetation categorisation, species identification, and the evaluation of vegetation health indices can all benefit from the addition of optical data, which provides extra spectral information.

The dynamics, structure, and health of wetland vegetation can be better understood through regular monitoring with microwave remote sensing. It aids in the study of wetland ecosystems, the evaluation of habitat quality, and the promotion of wetland management and conservation (Pu 2017; Rees 1999) .

Wetland Biomass Estimation Wetland biomass may be estimated via microwave remote sensing, which is useful for calculating carbon stocks and measuring productivity. Longer-wavelength microwaves are sensitive to the water content of vegetation, enabling biomass estimates to be made. Biomass retrieval methods can be developed by calibrating the relationships between microwave backscatter and biomass using data collected in the field.

Microwave remote sensing can be used to estimate the biomass of wetland vegetation by analysing data from SAR. Because of its capacity to see through vegetation canopies and provide details about their structure and scattering properties, microwave remote sensing is well-suited for biomass estimate.

Here are the key steps involved in wetland biomass estimation using microwave remote sensing:

Data Acquisition: Collect SAR information from spacecraft like Sentinel-1 and RADARSAT. Single (HH) or dual (HH/VV or HV) polarisation microwave imagery is available from SAR sensors. Because it provides more information about the scattering properties of vegetation, dual-polarisation data is frequently employed for biomass estimation.

Preprocessing: The SAR data should be preprocessed to get rid of noise and fix sensor-specific abnormalities. Radiometric calibration, speckle filtering to mitigate speckle in images, and geometric correction for pinpoint precision in georeferencing are all part of the process.

Backscatter Analysis: We can estimate marsh biomass by analysing the SAR data's backscatter properties. The scattering of microwaves is significantly affected by the biomass of vegetation. In general, more scattering causes a greater intensity of backscatter when the biomass is large. Therefore, marsh biomass can be measured by the intensity of the backscatter.

Biomass Estimation Models: Create or use methods for biomass estimation that take into account the correlation between backscatter intensity and marsh biomass. Field observations of biomass and associated SAR data are often used in the development of these models. Backscatter intensity can be related to biomass through the use of regression techniques like multiple linear regression or machine learning algorithms.

Ancillary Data Integration: If you want your biomass estimates to be more precise, you should include additional data like field measurements of biomass, vegetation indicators, and species-specific details. When it comes to calibrating and validating biomass models, data gathered from field observations is the gold standard. SAR data can be supplemented with optical remote sensing data, in the form of vegetation indices, to better capture the aforementioned variability.

Calibration and Validation: Use field data or measurements taken on the ground to calibrate the biomass estimating models. Relationships between the backscatter intensity acquired from remote sensing and the actual biomass on the ground are established through disruptive harvesting or biomass sampling. The precision and consistency of the biomass estimates can be evaluated with the help of validation methods like cross-validation or independent field data.

Biomass Mapping: Use the validated biomass estimation models to create biomass maps of wetlands. Use the models with the SAR data to generate pixel-by-pixel biomass estimates. The generated biomass maps show where wetland biomass is located and how much it varies across the research area.

Temporal Analysis: Multiple SAR pictures captured at different times can be used to examine the seasonal changes in marsh biomass. Seasonal dynamics of wetland biomass, such as vegetation growth, phenology, or responses to environmental changes, can be better understood with the aid of temporal analysis. Wetland biomass changes over time can be monitored using time series analysis methods including biomass change detection and biomass productivity analysis.

Microwave remote sensing can be used to estimate wetland biomass, which is important for determining carbon stocks, learning about ecosystem dynamics, and bolstering wetland management and conservation efforts.

Wetland Permafrost and Peatland Studies Permafrost and peatland ecosystems can be studied with microwave remote sensing. Permafrost extent, active layer thickness, and peatland features can all be gleaned from microwave measurements of the uppermost layers of the ground. These findings are useful for assessing how susceptible wetland ecosystems are to the effects of global warming and melting permafrost.

SAR data is used in studies of wetland permafrost and peatlands to investigate the properties and dynamics of permafrost and peatland habitats. Since microwaves can travel through layers of flora and soil to reveal qualities underneath, this method is ideal for these investigations.

Here are the key applications of microwave remote sensing in wetland permafrost and peatland studies:

Permafrost Mapping: Permafrost in wetland regions can be mapped using microwave remote sensing. The frozen state of permafrost is a defining feature that alters the dielectric characteristics of the subsoil. SAR data, especially at longer wavelengths (such as L-band), can provide information regarding the presence or absence of permafrost by penetrating the active layer. Permafrost borders can be determined by measuring the difference in backscatter intensity and coherence between frozen and thawed ground.

Active Layer Thickness Estimation: Wetland permafrost dynamics are highly dependent on the active layer, the layer of the soil that thaws seasonally above the permafrost. The active layer's thickness can be estimated by microwave remote sensing. Changes in active layer thickness can be observed by studying time series of backscatter intensity and coherence. Wetland permafrost habitats are particularly vulnerable to the effects of climate change, and the active layer thickness estimation can shed light on both of these phenomena.

Peatland Classification: Microwave remote sensing can be useful for categorising peatland ecosystems. Peatlands have a high percentage of water and organic matter in their soil. The specific backscatter characteristics of peatlands allow SAR data to distinguish them from other land cover types. Peatland areas can be identified and mapped by evaluating the backscatter intensity and polarisation characteristics, which aids in wetland management and conservation.

Peatland Subsurface Characterisation: The subsurface features of peatlands can be gleaned using microwave remote sensing. The density, amount of organic matter, and moisture level of the peat soil all play a role in how the microwaves interact with the peat. Peat parameters like moisture content, organic carbon content, and peat density can be estimated by studying the backscatter features, such as polarisation and frequency-dependent behaviour. These estimates improve our comprehension of hydrological processes in peatlands and the dynamics of carbon inside them.

Wetland Thaw Subsidence Monitoring: Wetland surfaces may sink or settle due to thaw subsidence as permafrost thaws. Wetland permafrost zones can be monitored for melt subsidence using microwave remote sensing. Subsidence rates can be determined by studying the time series of backscatter intensity or interferometric SAR data (InSAR). Wetland ecosystem dynamics and infrastructure design in permafrost regions rely on accurate measurements of thaw subsidence.

Wetland Carbon Monitoring: In permafrost and peatland regions, microwave remote sensing can aid in wetland carbon monitoring. Carbon is stored in large quantities in carbon-rich wetland ecosystems like peatlands. It is possible to assess carbon stocks and track changes in wetland carbon content by examining the backscatter features, such as polarisation and frequency-dependent behaviour. This data is helpful for carbon accounting, climate change research, and protecting wetlands.

Permafrost mapping, active layer thickness estimation, peatland classification, subsurface characterisation, thaw subsidence monitoring, and wetland carbon monitoring are just a few examples of the useful information that can be gleaned from microwave remote sensing of wetland and peatland environments. These uses further our knowledge of wetland dynamics and the effects of climate change, and they aid in the efficient administration and preservation of these special habitats.

Wetland Fire Detection and Monitoring Wetland fires can be detected and tracked with the help of microwave remote sensing. Since microwaves can see through smoke and clouds, they can detect fires that are otherwise hidden from the naked eye. To detect and track wetland fires, SAR data can be processed with active fire detection algorithms.

Microwave remote sensing can be used to identify, map, and monitor wetland fires, making it a useful tool for early detection. The scope, intensity, and behaviour of the fire can all be gleaned using microwave remote sensing.

Here are the key applications of microwave remote sensing in wetland fire detection and monitoring:

Fire Detection: Even in difficult situations like smoke or cloud cover, fires in wetland areas can be detected using microwave remote sensing. To detect the variations in backscatter strength generated by the fire, SAR data can be seen through air interferences. Fire hotspots can be identified and mapped by the study of temporal variations in backscatter. Using SAR data and fire detection algorithms, damaged areas can be automatically pinpointed.

Fire Mapping: Once a fire has been discovered in a wetland area, microwave remote sensing can be utilised to create a map of the fire's borders and extent. High-resolution images using SAR data enable for precise mapping of burned areas. Scars and burned areas can be mapped by comparing the intensity and coherence of backscatter before and after a fire. These maps are essential for determining the size of the fire and directing suppression and management activities.

Fire Behaviour Analysis: Wetland fire dynamics and behaviour can be better understood with the help of microwave remote sensing. Fire properties including front speed, spread direction, and intensity can be estimated by studying the temporal variations in backscatter intensity. These factors contribute to a better grasp of the fire's behaviour and dynamics, which in turn aids in the development of effective control techniques and protects firefighters.

Burn Severity Assessment: Wetland ecosystem damage and fire severity can be evaluated with microwave remote sensing. The extent of the burn can be estimated by comparing SAR data acquired before and after the fire, specifically by looking at the changes in backscatter intensity and coherence. The most impacted areas can be pinpointed with the help of burn severity maps, which can then be used in post-fire recovery evaluations and to direct restoration activities.

Post-Fire Monitoring: Post-fire wetland monitoring can benefit from the use of microwave remote sensing. After a fire has devastated a wetland's ecosystem, scientists can track its regrowth by collecting SAR data at regular intervals. Vegetation regrowth and recovery processes may be indicated by variations in backscatter intensity and coherence over time. This data is useful for determining the fire's long-term effects and developing post-fire management and restoration plans.

Integration with Other Data Sources: Wetland fire detection and monitoring can be improved by combining microwave remote sensing with additional data sources like optical remote sensing or weather data. Fire hotspots, smoke plumes, and the dynamics of vegetation can all benefit from the addition of optical data.

Contextual information on fire behaviour and fire danger circumstances can be gleaned from meteorological data such as temperature, humidity, and wind speed.

Early fire detection, precise mapping, and an evaluation of fire behaviour are all aided by microwave remote sensing for wetland environments. Wetland ecosystems and biodiversity can be better protected, and fire management can be improved, with the use of this data.

Wetland Subsidence and Erosion Monitoring Wetland subsidence and erosion may be tracked with great accuracy using microwave remote sensing. Wetland topographic changes, like as subsidence or erosion, can be detected by comparing SAR photos obtained at different times. These details are useful for comprehending the dynamics of wetland landscapes and the potential effects on wetland habitats.

5.2.1.1 Wetland Subsidence and Erosion Monitoring Using Microwave Remote Sensing

SAR data are used to monitor wetland subsidence and erosion using microwave remote sensing. This allows for the assessment of changes in wetland surface elevation and the detection of erosion trends over time. Since microwave remote sensing can see through plants to reveal subsurface features, it is ideal for these investigations.

Here are the key applications of microwave remote sensing in wetland subsidence and erosion monitoring:

Subsidence Mapping: Subsidence in wetland areas can be mapped and monitored using microwave remote sensing. When land gradually sinks or settles, this phenomenon is known as subsidence. Longer wavelength SAR data (such as L-band or P-band) can see through the plants and reveal changes in the ground's elevation. Subsidence can be detected and mapped by comparing SAR photos taken at varying periods of time.

Erosion Detection: Wetland erosion patterns can be seen and tracked with the aid of microwave remote sensing. Soil and other topsoil can be eroded away by wind and water. Erosion processes can be detected in SAR data through variations in backscatter intensity and coherence. Erosion hotspots can be located and their rate of spread monitored by studying the time series of SAR pictures.

Erosion Rate Estimation: Erosion rates in wetlands can be estimated using microwave remote sensing. Quantifying the quantity of material lost to erosion can be done by measuring the changes in backscatter intensity and coherence. Wetland areas' susceptibility to erosion can be better understood and erosion control and management techniques can be better implemented with accurate estimates of erosion rates.

Land Surface Deformation Monitoring: Surface changes in wetland areas can be tracked using microwave remote sensing. The topography can be measured in both vertical and horizontal directions. SAR data can be used with InSAR techniques like differential SAR interferometry (DInSAR) or persistent scatterer interferometry (PSI) to measure small elevation or displacement changes on the surface. Supporting both wetland management and infrastructure development, these methods provide precise assessments of land surface deformations such as subsidence or horizontal movements.

Wetland Change Detection: Large-scale wetland changes, like as subsidence and erosion, can be detected and monitored with the help of microwave remote sensing. It is feasible to evaluate the overall dynamics of wetland ecosystems by comparing SAR pictures obtained at different time periods and identifying areas undergoing substantial changes. SAR data can be used with change detection methods like image differencing or image thresholding to identify locations where there have been significant modifications.

Integration with Other Data Sources: Wetland subsidence and erosion monitoring can be improved by combining microwave remote sensing with other data sources like optical remote sensing or ground-based measurements. Changes in land cover, the dynamics of vegetation, and the quality of water are only some of the topics that can benefit from optical data. Validation and calibration data for the remote sensing measures can be obtained via ground-based measurements like surveying or global navigation satellite systems (GNSS).

The stability and variations of wetland areas can be better understood with the help of microwave remote sensing monitoring of subsidence and erosion. Wetland ecosystems and associated communities can be protected from the effects of subsidence and erosion with the use of this data, which will aid in wetland management, infrastructure development, and conservation initiatives.

The information gleaned from optical remote sensing is supplemented by the special capabilities of microwave remote sensing for studying wetlands. Microwaves are ideal for monitoring wetland hydrology, soil moisture, biomass, and other important features since they can pass through vegetation and clouds.

5.3 Hyperspectral Remote Sensing

The ability to gather and analyse data about the Earth's surface and atmosphere has been substantially improved with the advent of hyperspectral remote sensing, a cutting-edge and innovative technology. It is a major improvement over older remote sensing techniques like multispectral remote sensing because of the superior spectral resolution it offers. By measuring electromagnetic radiation in hundreds of discrete spectral bands, this technique provides a profoundly detailed picture of the surrounding world.

One of the primary advantages of hyperspectral remote sensing is its capacity to record and distinguish small fluctuations in the reflected or emitted electromagnetic radiation. Scientists may use this precise spectrum data to accurately classify materials, gases, and environmental aspects. Hyperspectral sensors are an important part of precision farming because of their ability to detect disease and pests, as well as pinpoint crop health. Hyperspectral data is useful in mineral prospecting because it may be used to determine the presence and location of individual minerals.

Hyperspectral remote sensing is also extremely useful for keeping tabs on the environment and protecting wildlife. The detection and analysis of the spread of invasive species, the condition of forests, and the presence of toxins and pollutants in water are all within its capabilities. This technology plays a crucial role in disaster management since it enables a quick evaluation of damage following events like floods, earthquakes, and wildfires. As an added bonus, hyperspectral remote sensing can be used in urban planning to evaluate the urban heat island effect and categorise land cover types. In ecological research, it is useful for assessing biodiversity, ecosystem health, and habitat quality.

The great spectral resolution of hyperspectral data makes it a significant tool in atmospheric study. It can be used to examine air quality, track changes in atmospheric composition, and study greenhouse gas emissions. This is of paramount importance in environmental research and policymaking about climate change.

Hyperspectral remote sensing is a game-changing tool for a wide variety of fields, including but not limited to agriculture; geology; environmental monitoring; disaster management; urban planning; ecology; and atmospheric research. Its ability to produce precise spectrum data has made it a crucial instrument for studying the Earth's surface and its evolving ecology, which has resulted in better resource management and more well-informed decisions. Hyperspectral remote sensing has already been a game-changer for our knowledge of Earth's intricate systems, and it's just going to get bigger as technology improves (Richards 2009; Sabins Jr and Ellis 2020; Shokr and Sinha 2023).

5.3.1 Use of Hyperspectral Remote Sensing for Wetland Studies

Due to its ability to record and analyse spectrum data, hyperspectral remote sensing has shown to be an effective method for studying wetlands. Important applications of hyperspectral remote sensing in wetland studies include the following which are also diagrammatically represented in Fig. 5.3 and some examples in Table 5.5 for wetlands and in Table 5.6 for saline wetlands.

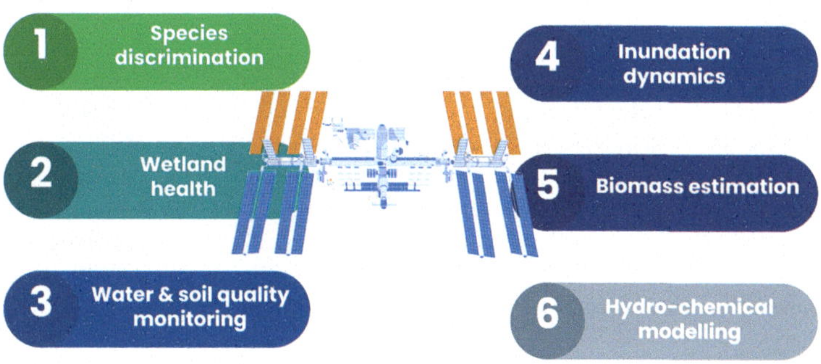

Fig. 5.3 Use of hyperspectral remote sensing for wetland studies

Table 5.5 Examples of hyperspectral remote sensing satellites for wetland use

S. no.	Name of wetland	Type of wetland	Hyperspectral satellite	Purpose of use
1	Everglades	Marshland	Hyperion	Vegetation and water quality assessment
2	Okavango Delta	Inland Delta	Environmental Mapping and Analysis Program (EnMAP)	Biodiversity monitoring
3	Pantanal	Seasonal Floodplain	CHRIS-PROBA	Wetland vegetation analysis
4	Sundarbans	Mangrove Forest	Airborne Visible InfraRed Imaging Spectrometer—Next Generation (AVIRIS-NG)	Mangrove cover change detection
5	Danube Delta	River Delta	HyMap	Ecosystem health assessment
6	Kakadu National Park	Wetland	AISA Eagle	Vegetation mapping
7	Hula Valley	Marshland	PRecursore IperSpettrale della Missione Applicativa (PRISMA)	Water quality and land use analysis
8	Tonle Sap Lake	Flooded Forest	Compact Airborne Spectrographic Imager (CASI)	Monitoring water quality and habitat
9	Lake Chad	Endorheic Lake	AISA Eagle	Mapping lake dynamics and shoreline
10	Okefenokee Swamp	Swamp	AISA Eagle	Studying wetland vegetation dynamics
11	Liaohe Delta	River Delta	CHRIS-PROBA	Monitoring deltaic land dynamics
12	Narew National Park	Wetland	HyMap	Tracking wetland water levels
13	Tisza River Basin	River Basin	PRISM	Monitoring river basin dynamics
14	Morava River Floodplain	Floodplain	EnMAP	Studying floodplain changes
15	Esteros del Ibera	Wetland	AISA Eagle	Mapping wetland changes
16	Poyang Lake	Seasonal Lake	CASI	Monitoring lake water level changes
17	Rann of Kutch	Seasonal Salt Marsh	Hyperion	Studying salt marsh dynamics
18	Itaipu Dam Reservoir	Reservoir	HyMap	Monitoring reservoir water levels
19	Humber Estuary	Estuary	PRISM	Mapping estuary water dynamics

Table 5.6 Examples of hyperspectral remote sensing for saline wetland

S. no.	Name of saline wetland	Type of saline wetland	Hyperspectral satellite used	Purpose of use
1	Chott el Jerid	Salt Flat	Earth Observation (EO)-1 Hyperion	Studying salt flat surface and minerals
2	Sambhar Lake	Salt Lake	EnMAP	Mapping lake water quality and sediments
3	Great Salt Lake	Terminal Lake	WorldView-3	Assessing lake water constituents
4	Lake Urmia	Endorheic Lake	AVIRIS	Monitoring lake conditions and salt content
5	Maranjab Salt Desert	Salt Desert	Hyperion	Mapping salt desert features and minerals
6	Etosha Pan	Salt Pan	Sentinel-2	Assessing pan conditions and wildlife
7	Lake Natron	Alkaline Lake	PRISMA	Mapping lake chemistry and shoreline
8	Salar de Uyuni	Salt Flat	AISA Eagle	Studying salt flat topography and minerals
9	Lake Eyre	Salt Pan	CASI	Mapping pan features and water quality
10	Lop Nur	Salt Pan	HyMAP	Monitoring salt pan conditions and topography
11	Badwater Basin	Salt Pan	Riegl LMS-Q780	Assessing salt pan topography and changes
12	Lake Abert	Endorheic Lake	Specim AISA	Monitoring lake water quality and sediments
13	Lake Tuz	Salt Lake	AISA Eagle	Mapping lake chemistry and conditions
14	Koyashskoye Salt Lake	Salt Lake	PRISMA	Assessing salt lake chemistry and minerals
15	Sebkha Sejoumi	Salt Pan	EnMAP	Studying salt pan features and composition
16	Lake Tengiz	Salt Lake	Specim AISA	Monitoring lake water chemistry and sediments
17	Lake Qarun	Salt Lake	WorldView-3	Mapping lake conditions and shoreline
18	Kara-Bogaz-Gol	Lagoon	PRISMA	Assessing lagoon water quality and features
19	Soda Lakes of Kenya	Saline Lakes	HyMAP	Studying saline lake chemistry and sediments
20	Lake Assal	Crater Lake	AISA Eagle	Mapping crater lake chemistry and sediments

5.3.1.1 Vegetation Mapping

Wetland vegetation types can be identified by their unique spectral signatures, which can be captured using hyperspectral sensors. Researchers can properly categorise and map wetland vegetation communities, such as emergent plants, submerged aquatic vegetation, and floating vegetation, by studying the reflected light over several narrow spectral bands.

Hyperspectral remote sensing is used for vegetation mapping by analysing the spectral features of vegetation over many small spectral bands using hyperspectral sensors. This method can be used to precisely categorise and map out various species of vegetation based on their characteristic spectral fingerprints.

Here's an overview of the process:

Data Acquisition: Airborne or satellite-based sensors capture the sunlight reflected from the Earth's surface in hundreds of narrow and contiguous spectrum bands to produce hyperspectral data. In the electromagnetic spectrum, different bands correspond to different wavelength ranges.

Preprocessing: Hyperspectral data undergoes extensive preprocessing to eliminate the impact of atmospheric conditions, compensate for sensor artefacts, and improve overall quality. Radiation-to-atmosphere-geometric calibration is the process involved.

Spectral Analysis: Spectral analysis is performed on the hyperspectral data after it has been prepared. This is done by inspecting the reflectance levels in all the various spectral bands for each pixel. Different types of vegetation have distinct spectral fingerprints, or patterns of reflectance values.

Training Data Collection: To train the classification algorithms, ground truth data must be gathered in the field. Spectral measurements must be taken of representative samples of the various types of plants found in the research area.

Classification: The hyperspectral data is used as a training dataset for supervised and unsupervised classification systems. Pixels with comparable spectral features are grouped into various classes in unsupervised classification, while in supervised classification, known vegetation types are assigned to the spectral signatures.

Accuracy Assessment: By comparing the categorised image to ground truth data gathered in the field, the accuracy of the vegetation classification results may be determined. This aids in assessing the reliability of the vegetation mapping outcomes.

Mapping and Visualisation: The final stage is to create vegetation maps from the categorised data. These maps show the abundance and distribution of several plant communities across the research region. The many types of vegetation are typically visualised using tools like false-colour composites and thematic maps.

Land cover mapping, habitat evaluation, ecosystem monitoring, and environmental management are just few of the uses for the data gleaned through hyperspectral remote sensing's application to vegetation mapping.

5.3.1.2 Species Discrimination

Hyperspectral information can be utilised to identify and separate a wide variety of wetland plant species. Because different species have different spectral properties, it is possible to identify and keep tabs on individual marsh plant species. Assessments of biodiversity, detection of invasive species, and monitoring of habitats can all benefit greatly from this data.

Species Discrimination Using Hyperspectral Remote Sensing

Spectral signatures collected by hyperspectral sensors allow for the identification and distinction of specific plant species, allowing for species discrimination via remote sensing. An outline of the steps is provided below:

Data Acquisition: Airborne or satellite-based sensors capture the sunlight reflected from the Earth's surface in hundreds of narrow and contiguous spectrum bands to produce hyperspectral data. In the electromagnetic spectrum, different bands correspond to different wavelength ranges.

Preprocessing: To improve the quality of the collected hyperspectral data and get rid of any sensor artefacts, the data goes through a series of preprocessing stages that include radiometric calibration, atmospheric correction, and geometry correction.

Spectral Analysis: The spectral signatures of several plant species are extracted from the preprocessed hyperspectral data. This is done by inspecting the values of reflectance in various spectral bands for each pixel representing a plant in a picture.

Training Data Collection: To train the classification algorithms, ground truth data must be gathered in the field. The first step is to determine which plant species are of interest, and then collect spectral data from representative samples of those plants.

Classification: The hyperspectral data is fed into supervised classification algorithms with the help of the training dataset. These formulas correlate spectral signals with identified plant species. Maximum likelihood, support vector machines, and decision trees are all examples of popular categorisation methods.

Accuracy Assessment: By comparing the categorised image to the ground truth data acquired in the field, the accuracy of the species discrimination results may be evaluated. This aids in assessing the reliability of the species identifications as a whole.

Mapping and Visualisation: The final step is to create maps or photographs that show how various plant species are dispersed across the area under investigation. The many types of species can be visualised in a way that's easy to understand with tools like false-colour composites and thematic maps.

Notably, a strong training dataset that includes a representative range of spectral signatures for each target species is essential for successful species discrimination utilising hyperspectral remote sensing. Accurate species discrimination requires taking into consideration a number of elements that can affect spectral fingerprints, including phenological fluctuations, canopy structure, and environmental conditions.

Hyperspectral remote sensing can be used for species discrimination, which has several uses beyond just species monitoring and invasive species detection. It's a great resource for learning about the make-up, distribution, and temporal shifts of different types of plant communities.

5.3.1.3 Wetland Health Assessment

Hyperspectral remote sensing can be used to assess wetland health and condition by gauging indicators of plant stress like chlorophyll content, leaf water content, and pigment concentrations. These markers can be used to identify environmental changes in wetland ecosystems, such as those caused by pollution, invading species, or changing hydrology.

Wetland Health Assessment Using Hyperspectral Remote Sensing

Wetland ecosystem health can be assessed with hyperspectral remote sensing by looking at a variety of spectral indications in the data. Below is a summary of the procedures:

Data Acquisition: Airborne or satellite-based sensors capture the sunlight reflected from the Earth's surface in a wide range of discrete spectral bands to produce hyperspectral data. In the electromagnetic spectrum, different bands correspond to different wavelength ranges.

Preprocessing: Preprocessing steps including radiometric calibration, atmospheric correction, and geometry correction are used to enhance the quality of the obtained hyperspectral data and remove any sensor artifacts.

Spectral Analysis: The preprocessed hyperspectral data is then used to extract spectral markers of wetland health. These indicators can take the form of indices of vegetation, concentrations of pigments, chlorophyll fluorescence, water content, or any number of other biophysical factors.

Calibration and Validation: The spectral indicators calculated from hyperspectral data are calibrated and verified using ground-truth measurements and field observations. This will make it easier to correlate spectral data with indicators of wetland health.

Wetland Health Parameters: Wetland health indicators that can be derived from hyperspectral data include chlorophyll concentration, leaf water content, photosynthetic activity, stress in vegetation, nutrient concentrations, and vegetation type. From these numbers, we may infer a lot about the health of wetland ecosystems and the vegetation that thrives there.

Change Detection: By comparing hyperspectral data obtained at different dates, changes in wetland health features can be determined. Pollution, invasive species, and hydrological changes are just some of the hazards that can be identified and addressed in this way to protect wetland ecosystems.

Mapping and Visualisation: Data from wetland health evaluations can be visualised through the use of spatial or themed maps. Wetland health parameters are visualised on these maps to identify hotspots for issues, decline, and restoration opportunities.

Researchers and environmental managers can employ hyperspectral remote sensing as part of a wetland health assessment to efficiently and effectively monitor the state and dynamics of wetland ecosystems. Using this information, we can better understand wetland biodiversity, monitor ecosystem health, focus conservation efforts, and promote sustainable wetland management practises (Thenkabail and Lyon 2016; Toro and Tsourdos 2018).

5.3.1.4 Water Quality Monitoring

Parameters of water quality that can be measured using hyperspectral sensors include turbidity, suspended sediments, and dissolved organic matter concentrations. Researchers can evaluate the wetland ecosystem's water quality as a whole, as well as detect any changes or pollution events, by studying the light's absorption and scattering qualities in the water.

Hyperspectral remote sensing is used for water quality monitoring, in which a variety of water quality parameters are evaluated and analysed using hyperspectral data. Water quality factors can be better understood with the help of hyperspectral remote sensing, which analyses the spectral signatures of water bodies at various wavelengths. The steps involved are as follows.

Data Acquisition: Airborne or satellite-based sensors capture the sunlight reflected from the Earth's surface in a wide range of discrete spectral bands to produce hyperspectral data. In the electromagnetic spectrum, different bands correspond to different wavelength ranges.

Preprocessing: To improve the quality of the collected hyperspectral data and get rid of any sensor artefacts, the data goes through a series of preprocessing stages that include radiometric calibration, atmospheric correction, and geometry correction.

Spectral Analysis: Hyperspectral data is processed beforehand and then examined to derive water quality-related spectral information. This is done by inspecting the levels of reflectance in various spectral bands for each pixel representing a body of water in an image.

Water Quality Parameters: Hyperspectral data can be used to infer many characteristics of water quality, like as.

Chlorophyll-a concentration: Chlorophyll-a can be used to evaluate eutrophication and algal blooms in water bodies because it is a measure of algal biomass.

Total Suspended Solids (TSS): TSS concentration is a measure of sedimentation and water clarity.

Dissolved Organic Matter (DOM): DOM is a useful indicator of the amount of organic chemicals in water that contribute to its colour, turbidity, and biological activity.

Water temperature: The biological activities, thermal stratification, and ecosystem health are all affected by the water temperature.

Water depth: Water depth can be estimated using hyperspectral data by examining the light attenuation with depth.

Water clarity: The Secchi disc depth is a standard way to quantify water clarity, and it may be estimated with hyperspectral data.

Calibration and Validation: Ground-truth measurements and in-field observations are used to calibrate and verify the spectral information obtained from hyperspectral data. This ensures a reliable evaluation by establishing a connection between spectral data and water quality metrics.

Mapping and Visualisation: Thematic maps and other spatial representations can be used to display the findings of water quality monitoring. These maps show where water quality parameters are concentrated, which can help pinpoint problem regions, pollution sources, or trends in water quality.

Hyperspectral remote sensing for monitoring water quality is an effective method for evaluating and controlling water supplies. It helps in environmental monitoring, water resource management, and decision-making by allowing for the identification of changes, pollution events, and the spatial distribution of water quality metrics over broad areas.

5.3.1.5 Wetland Hydrology

Wetland hydrological processes such as water flow patterns, inundation dynamics, and surface water-vegetation interactions can be studied with hyperspectral data. Hydrological cycles, water storage capacity, and the effects of climate change and land use activities on wetland hydrology can all be better comprehended with this data in hand.

The use of hyperspectral data to explore different elements of wetland hydrological processes and interactions between water and vegetation is what constitutes a wetland hydrology study utilising hyperspectral remote sensing. What follows is a brief summary of the benefits that hyperspectral remote sensing can bring to the study of wetland hydrology:

Water Flow Patterns: Water flow patterns in wetlands can be analysed with the aid of hyperspectral remote sensing. Scientists can learn about the dynamics of water circulation in wetland systems by monitoring temporal shifts in the spectral characteristics of water bodies.

Inundation Dynamics: Inundation of wetlands can be tracked and analysed in real time and across large areas with the help of hyperspectral data. Wetland flooding can be detected, measured, and monitored over time by analysing the spectral fingerprints of water bodies and surrounding plants. The hydrological cycles of wetlands and how they react to changes in climate and land use can be better understood using this data.

Surface Water-Vegetation Interactions: Hyperspectral remote sensing is useful for studying how water and vegetation in wetlands interact with one another. Wetland plant communities' vitality, distribution, and health can be evaluated via the lens of water availability, inundation duration, and water quality by analysing the spectral properties of vegetation in various wetland locations. Hydrological processes and the dynamic nature of plants in wetland ecosystems can be better comprehended with the aid of this data.

Water Storage Capacity: Wetlands' potential to store water can be better understood with the help of hyperspectral data. Researchers can estimate the water volume stored inside wetlands and track oscillations in water storage capacity by monitoring changes in the spectral features of water bodies throughout wet and dry times. This data helps manage water supplies and determine how resistant wetlands are to drought and flooding.

Hydrological Modelling and Mapping: Wetland hydrological simulations can be made more accurate with the help of hyperspectral data when incorporated into hydrological models. Researchers can improve their estimates of water flow, evapotranspiration, and water balance in wetland systems by including spectral information like water reflectance and vegetation indices in modelling frameworks. It is also possible to create hydrological maps using hyperspectral data, which show the regional variations in wetland hydrological properties such as water depth, water table levels, and soil moisture content.

Hyperspectral remote sensing investigations of wetland hydrology shed light on the intricate water dynamics of wetland ecosystems. Wetland hydrological processes, water resource management, conservation planning, and the evaluation of wetland health and resilience in the face of environmental changes can all benefit from this knowledge.

Wetland Change Detection: Changes in wetland size, vegetation cover, and water quality can all be tracked and detected by comparing hyperspectral data collected at different times. This helps conservation and management efforts by pinpointing instances of wetland loss, degradation, or restoration.

5.3.1.6 Wetland Change Detection Using Hyperspectral Remote Sensing

Hyperspectral remote sensing can be used to detect changes in wetland ecosystems through the analysis of hyperspectral data collected at many time points. Changes in wetland extent, vegetation cover, water quality, and other factors can be detected and evaluated by comparing the spectral features of wetland areas across time. The steps involved are as follows:

Data Acquisition: Hyperspectral data captures the sunlight reflected from the Earth's surface in several narrow and continuous spectral bands and is acquired using airborne or satellite sensors at multiple time periods.

Preprocessing: To improve the quality of the collected hyperspectral data and get rid of any sensor artefacts, the data goes through a series of preprocessing stages that include radiometric calibration, atmospheric correction, and geometry correction.

Spectral Comparison: Hyperspectral data, after being processed, is compared across time periods to spot shifts. This is done by assessing variances in reflectance values across spectral bands and looking at spectral signatures of wetland areas.

Change Detection Algorithms: In order to identify and categorise changes, the results of the spectral comparisons can be fed into a number of different change detection algorithms. Differential image processing, principal component analysis, change vector analysis, and spectral mixture analysis are only few of the algorithms in this category. Using these methods, we can better see where the wetland ecology is changing the most.

Accuracy Assessment: Results from change detection are evaluated by comparing them to "ground truth" data or to observations made in the field. The accuracy and precision of the change detection procedure can then be assessed.

Change Mapping and Visualisation: Using mapping and visualisation, the results of wetland change detection may be effectively communicated. Thematic maps and spatial representations can be used to visualise changes in wetland ecosystems such as wetland loss, vegetation change, water quality degradation, and others.

Long-term monitoring and assessment of wetland ecosystems is made possible with the help of hyperspectral remote sensing. It may be used to better understand wetland ecosystems, detect when they are being lost or degraded, evaluate the success of conservation efforts, and guide land management and preservation choices.

When it comes to describing and keeping tabs on wetland ecosystems, hyperspectral remote sensing is an effective and non-intrusive method. Insights into wetland vegetation, water quality, and ecological processes are made possible by the instrument's high spectral resolution and precise spectral information, which contributes to the protection and sustainable management of these important habitats.

5.4 LiDAR Technology

Incredibly precise mapping and analysis of the Earth's surface and its features are now possible because of LiDAR remote sensing. It works by transmitting laser pulses from an airborne or ground platform and measuring the time it takes for the pulses to return after contact with a target, resulting in an accurate three-dimensional map. This technology has found multiple applications across a wide range of sectors, from geospatial and environmental science to infrastructure planning and catastrophe management.

LiDAR remote sensing's ability to collect precise elevation data at a high resolution is a major benefit. It's a staple in the field of topographic mapping, where it's used to generate accurate DEMs that are then put to use in applications as varied as hydrologic forecasting, urban development, and transportation planning. LiDAR data is used to create canopy height models and evaluate forest structure, both of which help with forest management and conservation.

LiDAR technology helps urban planners and civil engineers by allowing them to create highly accurate 3D models of urban environments. It makes it possible to evaluate buildings and transportation systems and spot risks. LiDAR is also extremely useful in archaeological exploration, as it can detect objects and buildings obscured by overgrowth or dirt, such as ancient ruins.

Monitoring and measuring vegetation, such as tree health, biomass estimation, and deforestation, is crucial in natural resource management, and LiDAR is an indispensable tool for this. The knowledge it provides about habitat structure and species distribution is crucial for ecosystem management. LiDAR has enormous potential uses in crisis management. It's useful for estimating disaster damages, detecting landslides, and modelling floods. It can promptly and precisely map areas affected by natural disasters, helping response operations and recovery.

In addition, LiDAR finds widespread application in the transportation and utility sectors. It's useful for coordinating the construction and upkeep of things like highways, railroads, and electricity lines. In general, the ability to produce highly detailed and precise 3D representations of the Earth's surface with LiDAR remote sensing is a game-changer in geospatial and environmental science. Its usefulness extends far beyond efforts to protect the natural world and includes disaster management, building infrastructure, and more. It is anticipated that LiDAR's importance will grow and that it will provide even deeper insights into our dynamic world as technology develops further. Key LIDAR uses in wetland research include the following which are also diagrammatically represented in Fig. 5.4 and some examples in Table 5.7 for wetlands and in Table 5.8 for saline wetlands.

LIDAR's topographic mapping capabilities extend to capturing the subtle details of wetland basins, channels, and floodplains. Wetland geomorphology can be better comprehended, and its boundaries better identified, with the help of high-resolution DEMs.

Laser pulses are used to detect distances and generate accurate elevation models of the Earth's surface in LIDAR topography mapping. In several contexts, LIDAR technology is used to create precise and thorough topographic maps.

The steps involved are outlined below:

Data Acquisition: Both aerial and ground-based LiDAR systems can be used to collect LiDAR data. LiDAR sensors can be fitted on aeroplanes or unmanned aerial vehicles to collect laser beams that fall to Earth. Stationary terrestrial LiDAR systems do a wide-area scan from a single vantage point. The LiDAR system fires out laser pulses, which bounce off the ground and nearby objects and are time-stamped as they return to the sensor.

Point Cloud Generation: The LiDAR data is used to generate a point cloud in three dimensions by processing the recorded laser returns. The X, Y, and Z coordinates of each point in the point cloud are the longitude, latitude, and elevation of that location on Earth.

Data Processing: A clean and precise picture of the Earth's naked surface is obtained by processing the point cloud data to eliminate noise, outliers, and vegetation points. To separate ground features from non-ground features like buildings and foliage, filtering methods are employed.

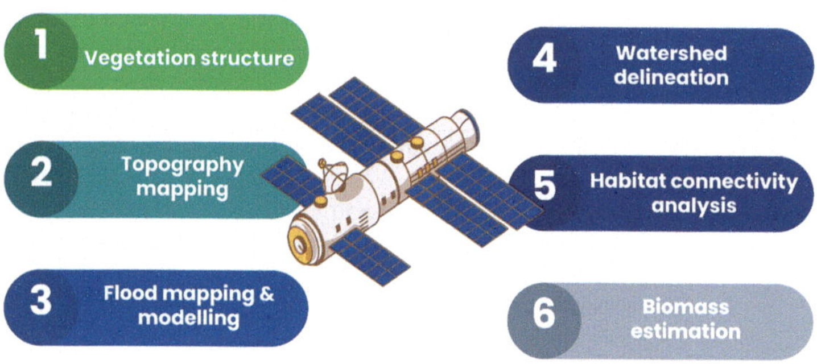

Fig. 5.4 Use of LiDAR for wetland studies

Table 5.7 Examples of LiDAR for wetland studies

S. no.	Name of wetland	Type of wetland	LiDAR remote sensing satellite	Purpose of use
1	Everglades	Marshland	Ice, Cloud, and land Elevation Satellite (ICESat-2)	Monitoring wetland water levels
2	Okavango Delta	Inland Delta	TerraSAR-X	Topographic mapping and vegetation
3	Pantanal	Seasonal Floodplain	ALSAT-1B	Flood modelling and land topography
4	Sundarbans	Mangrove Forest	GeoEYE-1	Topography and habitat assessment
5	Danube Delta	River Delta	PRISMA	Wetland ecosystem mapping
6	Kakadu National Park	Wetland	Riegl LMS-Q780	Vegetation and floodplain analysis
7	Hula Valley	Marshland	Leica ALS50	Water level and habitat monitoring
8	Tonle Sap Lake	Flooded Forest	Optech ILRIS-3D	Hydrology and habitat assessment
9	Lake Chad	Endorheic Lake	Velodyne HDL-32E	Lake level and shoreline changes
10	Okefenokee Swamp	Swamp	Leica ALS60	Vegetation and wetland topography
11	Liaohe Delta	River Delta	Riegl VZ-400	Deltaic landform and hydrology
12	Narew National Park	Wetland	Riegl VZ-400	Wetland health and water levels
13	Tisza River Basin	River Basin	Optech Orion	River basin topography and hydrology
14	Morava River Floodplain	Floodplain	Leica ALS50	Floodplain and river dynamics
15	Esteros del Ibera	Wetland	GeoEYE-1	Vegetation and wetland changes
16	Poyang Lake	Seasonal Lake	ICESat-2	Lake water level and shoreline
17	Rann of Kutch	Seasonal Salt Marsh	TerraSAR-X	Salt marsh dynamics and topography
18	Itaipu Dam Reservoir	Reservoir	PRISMA	Reservoir water level and topography
19	Humber Estuary	Estuary	Velodyne HDL-32E	Estuary topography and water levels
20	Salton Sea	Endorheic Lake	Leica ALS50	Lake level and habitat assessment

Table 5.8 Examples of LiDAR technology for saline wetland studies

S. no.	Name of saline wetland	Type of saline wetland	LiDAR remote sensing satellite	Purpose of use
1	Chott el Jerid	Salt Flat	ICESat-2	Monitoring salt flat topography
2	Sambhar Lake	Salt Lake	TerraSAR-X	Mapping lake and salt crust changes
3	Great Salt Lake	Terminal Lake	ALSAT-1B	Monitoring lake elevation and area
4	Lake Urmia	Endorheic Lake	GeoEYE-1	Assessing lake level changes
5	Maranjab Salt Desert	Salt Desert	ALOS-2	Mapping desert dune dynamics
6	Etosha Pan	Salt Pan	HERSAT	Monitoring salt pan features
7	Lake Natron	Alkaline Lake	LiDARLite V3	Studying lake topography and bird nests
8	Salar de Uyuni	Salt Flat	TerraScan	Assessing salt flat surface changes
9	Lake Eyre	Salt Pan	PRISM	Mapping lake and salt crust dynamics
10	Lop Nur	Salt Pan	Optech ILRIS-3D	Monitoring changes in salt pans
11	Badwater Basin	Salt Pan	ALSAT-1B	Measuring salt crust thickness
12	Lake Abert	Endorheic Lake	LMS-Z420i	Studying lake depth and water features
13	Lake Tuz	Salt Lake	Riegl LMS-Q780	Monitoring salt lake area and depth
14	Koyashskoye Salt Lake	Salt Lake	GeoEYE-1	Assessing salt lake water level changes
15	Sebkha Sejoumi	Salt Pan	Leica ALS60	Mapping salt pan topography
16	Lake Tengiz	Salt Lake	Optech Orion	Monitoring lake area and shoreline
17	Lake Qarun	Salt Lake	ALSAT-1B	Assessing lake level and shoreline
18	Kara-Bogaz-Gol	Lagoon	Velodyne HDL-32E	Studying lagoon dynamics
19	Soda Lakes of Kenya	Saline Lakes	Leica ALS50	Mapping saline lake features
20	Lake Assal	Crater Lake	Riegl VZ-400	Assessing crater lake topography

DEM Generation: A DEM or DSM (Digital Surface Model) can be made from the processed point cloud. While a DSM incorporates all topographical features, a DEM only portrays the bare Earth surface, without vegetation or man-made structures. All of the elevation data for the research area can be found in these models.

Contour Generation and Analysis: From the DEM, contour lines can be made to illustrate the landscape's topography. The landform and slope features of the area are depicted graphically by these contour lines, which are isolines linking points of equal height.

Terrain Analysis: The resulting topographic data can be studied further to glean details about the topography, such as its slope, aspect, flow direction, and flow accumulation. Applications such as infrastructure design, flood modelling, and land management can benefit from this research on drainage patterns, erosion potential, hydrological processes, and suitability assessments.

Visualisation and Mapping: The DEM, contour lines, and terrain features can all be viewed in GIS for use in spatial analysis and mapping thanks to the data's topographic nature. These maps are useful for a variety of geospatial purposes, including land use planning, engineering design, and environmental assessments.

In order to accurately portray the Earth's surface, LiDAR topography mapping delivers extremely accurate and detailed elevation data. It can be used for anything from precision agriculture and urban planning to hazard evaluation and environmental monitoring. Understanding and evaluating the properties and dynamics of the Earth's terrain is greatly aided by the high-resolution topographic information received using LiDAR technology.

5.4.1 Biomass Assessment

Using LiDAR, marsh vegetation structure may be accurately mapped. LiDAR data can estimate factors like vegetation height, canopy density, and vertical profiles by analysing the intensity of laser reflections from the vegetation canopy. This data is helpful for describing plant communities in wetlands, measuring biomass, and keeping tabs on how the vegetation structure is evolving over time.

The LiDAR method of assessing vegetation structure and biomass uses laser pulses to estimate the height, canopy density, and biomass of plants in three dimensions. In order to estimate biomass and enable numerous applications in forestry, ecology, and carbon accounting, LiDAR technology delivers thorough and reliable data on vegetation structure. The steps involved are as follows.

Data Acquisition: Data from LiDAR systems, either in the air or on the ground, are acquired. LiDAR sensors mounted on planes or unmanned aerial vehicles send out pulses of light that reflect off the tops of trees and are then recorded by the sensor.

Point Cloud Generation: The LiDAR data's recorded laser returns are utilised to generate a point cloud in three dimensions that shows the topography and spatial organisation of the plant life. Each dot in the point cloud represents a discrete spot in the overstory.

Canopy Height Model (CHM) Generation: CHMs can be calculated from a point cloud by subtracting the ground level from the height of the individual vegetation points. The CHM gives a two-dimensional map of vegetation canopy height and is a representation of the height of the plants above ground.

5.4.2 Vegetation Structure Analysis

Vegetation structure data is gleaned from the CHM. This requires calculating metrics like canopy mean height, canopy maximum height, canopy density, and vertical profiles of plants. The vertical distribution and arrangement of vegetation in the research area can be better understood with the use of these measurements.

Biomass Estimation: Vegetation biomass can be estimated using data collected from vegetation structure analysis and applied using biomass estimation techniques. The LiDAR metrics that are measured or computed are then used in biomass estimation algorithms. To estimate biomass, these models often take into account variables like canopy height, density, and diameter, however, this might vary by vegetation species and location.

Validation and Calibration: Validation and calibration are required to guarantee precision in biomass estimation. To verify the accuracy of the LiDAR-based biomass estimates, field measurements of vegetation biomass are gathered in strategic sites throughout the study region. By comparing LiDAR-derived metrics to field observations, we may calibrate the biomass estimation algorithms.

Mapping and Visualisation: Spatial representations of canopy height, biomass distribution, and other important factors can be derived from mapping and visualising the data on vegetation structure and biomass. These maps are useful for locating regions with abundant biomass, dense vegetation, or promising habitat quality.

The use of LiDAR technology for assessing vegetation structure and biomass is a quick and effective way to learn about and track changes in plant life. Forest management, carbon sequestration research, ecological study, and conservation planning all benefit from LiDAR's precise biomass estimates.

5.4.3 Hydrological Analysis

Wetland hydrology research can benefit from LiDAR-derived elevation data. LIDAR is useful for determining water volume, water storage capacity, and flow dynamics by precisely outlining water bodies, canals, and floodplains. This data is useful for studying wetland hydrological processes, modelling floods, and managing water supplies.

LiDAR technology is used in hydrological studies to investigate and examine a wide range of water-related processes and phenomena. In order to comprehend a landscape's topography, hydrology, and water dynamics, LiDAR's high-resolution elevation data is required. Here are some of LiDAR's most important uses in hydrological research:

Topographic Mapping: High-resolution DEMs that faithfully depict topography are generated using LIDAR data. Surface features such as lakes, rivers, and other bodies of water, as well as the inclinations and depressions of the ground, can all be determined with the use of these elevation models. Understanding water flow patterns, locating drainage networks, and studying watershed characteristics all require this information.

Floodplain Mapping and Flood Modelling: When it comes to charting and simulating floodplains, LiDAR-derived elevation data is important. LiDAR is useful for identifying flood-prone locations, determining floodplain extents, and assessing flood risks since it provides an accurate representation of the ground surface. Hydrological models can simulate and predict flood events with the help of elevation data, which helps with flood mitigation, emergency planning, and floodplain management.

Channel and Stream Morphology: River, stream, and channel morphology and features can be studied in great depth with the help of LiDAR data. It's useful for pinpointing channel beginnings and endings as well as channel profiles and cross-sections. This data is crucial for determining the integrity of river channels, evaluating sediment movement, and understanding stream hydraulics.

Watershed Delineation and Hydrological Modelling: Watershed boundaries and delineation are established using elevation data obtained from LiDAR. Hydrological modelling, which simulates processes like rainfall-runoff, streamflow routing, and water balance calculations, benefits from the precise topographic data generated from LiDAR. This modelling is useful for determining the effects of land use changes on hydrological processes, analysing water resource management techniques, and learning more about water availability.

Restoration of Rivers and Streams: LiDAR data is used in these efforts. Channel realignment, floodplain reconnection, and habitat restoration are all made easier with the use of high-resolution elevation models. If you want to know how well your restoration project turned out, you can use LiDAR to compare the landscape before and after it was worked on.

Wetland Mapping and Analysis: Wetland areas can be better mapped and analysed with the help of LiDAR data. In order to locate wetlands, describe their topography, and evaluate their hydrological dynamics, high-resolution elevation data is essential. Wetland depressions, canals, and water storage capabilities are all features that benefit greatly from being mapped with LiDAR.

Supporting water resource management, floodplain mapping, watershed planning, and ecological restoration, hydrological analysis employing LiDAR technology improves our understanding of water-related processes and characteristics. LiDAR's high-resolution elevation data meticulously depicts the landscape, allowing for in-depth and precise hydrological study.

5.4.4 Wetland Change Detection

By comparing LiDAR data from several time periods, we can learn about shifts in wetland size and shape. Researchers can pinpoint locations of wetland loss, expansion, or changes in topography by comparing different elevation models. This is useful for keeping an eye on wetland activity, measuring the effects of humans, and directing conservation and restoration efforts.

LiDAR can be used to detect and assess changes in wetland area, morphology, and vegetation cover by comparing LiDAR data gathered at different time points. Since LiDAR provides both precise topographic information and high-resolution elevation data, it can be used to detect and track changes in wetland ecosystems. Here's an overview of the process:

Data Acquisition: The wetland region of interest is captured by collecting LIDAR data at numerous time intervals. Airborne LiDAR sensors installed on planes or drones can accomplish this. To document the evolution of wetland features through time, researchers conduct repeated surveys of the same area.

Point Cloud Comparison: Point clouds, constructed from LiDAR data collected at various times, show the wetland's three-dimensional structure. Differences in topography and vegetation structure can then be picked out by comparing point clouds from several acquisitions.

DSM Comparison: After the LiDAR point clouds are processed, a DSM of the wetland's surface is generated. By comparing DSMs acquired at various times, changes in the wetland's topography can be identified. This includes, but is not limited to, changes in elevation, landform modifications, and the size and shape of water bodies.

Canopy Height Model (CHM) Comparison: By deducting the terrain elevation from the LiDAR-derived surface elevations, CHMs are created. The height and structure of vegetation can be gleaned from CHMs. Changes in vegetation cover, vegetation height, or alterations in wetland vegetation communities can be detected by comparing CHMs from different time points.

Change Detection Analysis: The contrasted datasets can be subjected to a number of change detection methods, including thresholding, image differencing, and change vector analysis. Changes in height, water extent, or vegetation cover can all be easily spotted thanks to these methods. Based on these criteria, the discovered changes can be classified and measured.

Interpretation and Validation: Ground-truth information, field surveys, and supplementary remote sensing data are used to evaluate and verify the changes that have been discovered. Wetland transitions can be attributed to a variety of factors, including shifts in land use, natural processes, and human impacts, all of which can be confirmed through field visits or ground-based observations.

Change Analysis and Reporting: Extent, shape, and vegetation cover shifts in wetlands are tracked and reported upon. Assessing the effects on wetland ecosystems and learning about the broader ecological ramifications requires quantifying the size and dispersion of the alterations.

Using LiDAR to identify changes in wetlands paves the way for more accurate assessments of wetland dynamics, health monitoring, and the assessment of management and restoration initiatives. LiDAR's precise elevation data helps decision-makers and conservationists map and analyse wetland changes in great detail, which benefits efforts to protect these vital ecosystems.

Habitat Assessment and Species Mapping: Wetland ecosystems can benefit from using LiDAR data for habitat evaluation and species mapping. Different types of habitats, such as open water, emergent vegetation, shrub zones, and forested areas, can be distinguished thanks to the detailed information about vegetation structure and terrain. Wildlife habitat models, species distribution maps, and preservation strategies can all benefit from these numbers.

5.4.5 Habitat Assessment

LIDAR is used for habitat evaluation when studying and analysing habitat features such as vegetation structure, topography, and landscape characteristics. LiDAR is useful for assessing and monitoring habitats because it gives high-resolution elevation data and accurate topography information. Here's an overview of how LiDAR can be used for habitat assessment:

Habitat Mapping and Classification: Different habitat types can be mapped and classified using structural properties using LiDAR data. Detailed elevation maps reveal the many landscape features, plant communities, and land use categories present in a given ecosystem. Forests, marshes, grasslands, and shrublands can all be delimited and categorised using LiDAR-derived data including elevation, canopy height, and vegetation density.

Vegetation Structure Analysis: Canopy height, density, and vertical profiles are only some of the details about vegetation structure that may be gleaned from LiDAR data. Researchers can evaluate the vertical and horizontal complexity of a habitat's vegetation by examining the LiDAR-derived metrics, such as the CHM. This research aids in the evaluation of habitat quality and fragmentation as well as the identification of wildlife corridors.

Wildlife Habitat Modelling: Wildlife habitat appropriateness can be evaluated using LiDAR data in conjunction with species distribution models. Researchers can develop predictive models of optimal habitat areas for target species by combining LiDAR-derived factors like terrain ruggedness, vegetation density, and canopy height with data on species occurrence or habitat preference. These models are useful for pinpointing ecologically significant areas, setting conservation priorities, and informing land use policy.

Habitat Connectivity Analysis: The assessment of habitat connectivity and landscape connectivity corridors is aided by LiDAR-derived elevation data. Researchers can find potential corridors for species to migrate across habitats by evaluating topographic characteristics including ridges, valleys, and water bodies. Understanding habitat fragmentation, assessing connectivity for wildlife populations, and pinpointing places in need of restoration or conservation are all aided by this data.

5.4.6 Wetland Assessment

Wetland habitat assessment is aided by LiDAR data, which details the size, shape, and composition of these environments. It's useful for locating wetlands, classifying what kinds of wetlands they are (such as marshes, swamps, or bogs), and gauging the way water moves through them. Understanding wetland inundation patterns, water flow dynamics, and temporal variations in wetland extent is further aided by LiDAR-derived elevation data.

Habitat Change Detection: Changes in habitat characteristics can be identified using LiDAR data collected at various times. Vegetation cover, canopy height, and topographic feature changes can all be detected by comparing LiDAR-derived information. This allows for a more accurate assessment of habitat degradation, evaluation of the effects of land use changes, and tracking of the progress of habitat restoration.

The structure, connectivity, and dynamics of ecosystems can be better understood by LIDAR-based habitat assessments. Supporting habitat protection, land management, and ecological study, LIDAR's high-resolution elevation data and accurate topography information allow for extensive analysis and mapping of ecosystems.

5.4.7 Wetland Restoration and Management

For wetland management and restoration efforts, LiDAR data is invaluable. It is useful for comparing elevation models before and after wetland restoration to determine whether or not the efforts were successful. Hydrological infrastructure, such as levees, dikes, or water control structures, can also benefit from LiDAR's use in their planning and construction.

LIDAR technology has considerable potential for use in wetland restoration and management. Wetland restoration planning, implementation, and monitoring are greatly aided by the precise elevation data and topographic information provided by LIDAR. Here's an overview of how LIDAR can be used in wetland restoration and management:

Baseline Assessment: Wetland evaluation and characterisation can begin with LiDAR data. It shows in detail the wetland's terrain, water levels, and plant community. Knowing the wetland's current state, pinpointing problem areas, and establishing restoration goals all benefit from a baseline survey.

Hydrological Analysis: The hydrological properties of wetlands can be better understood with the help of elevation data acquired from LiDAR. It's useful for pinpointing the locations of water sources, drainage systems, and flood plains. Wetland restoration projects can take advantage of opportunities to restore natural water flows, reconnect hydrological systems, and build water retention structures when hydrology is taken into account.

Wetland Mapping and Design: By giving precise details regarding wetland boundaries, vegetation cover, and landforms, LiDAR data aids in wetland mapping and planning. This is useful for defining wetlands, pinpointing regions in need of restoration, and creating functional wetlands with features like ponds, canals, or wetland cells. Wetland design designs benefit from the accuracy and specificity made possible by LiDAR-derived elevation data.

Water Storage and Management: Wetlands are important for storing water, and LiDAR data can assist in assessing their storage capacity. Researchers can predict how much water can be stored in different wetlands by evaluating elevation data. This data is helpful for assessing the water needs of wildlife and downstream habitats and developing more effective management plans.

Habitat Restoration and Monitoring: Restoration of wetland habitats can be better planned and executed with the help of LiDAR data. Wetland managers can use elevation data to pinpoint potential sites for new habitats, regrading existing wetlands, and replanting of native species. By comparing elevation data from before and after a restoration project, as well as evaluating changes in plant cover and habitat quality, LiDAR can be used to track the progress of these efforts.

Wetland Change Detection: Changes in wetland conditions can be detected and monitored using LiDAR data collected at various times. Changes in wetland area, water depths, or vegetation structure can all be detected by managers by comparing LiDAR-derived elevation data. This data is useful for monitoring the condition of wetlands, gauging the success of management efforts, and pinpointing problem areas.

5.4.8 Risk Assessment and Floodplain Management

Wetland risk assessment and floodplain management benefit from LiDAR-derived elevation data. Wetland managers can assess flood risks and take preventative action to lessen the damage floods cause to wetland ecosystems and nearby towns by assessing elevation data and floodplain extents.

Wetland restoration and management can benefit greatly from the data collected by LiDAR systems since it allows for more accurate planning, more efficient implementation, and better decision-making. LiDAR's precise topographic data and high-resolution elevation maps help us better understand wetland dynamics, protect vital wetland ecosystems, and ensure the long-term viability of restored wetlands.

In order to better understand wetland properties, dynamics, and functions, LiDAR technology provides detailed and accurate spatial information. Wetland assessment, monitoring, and conservation are greatly aided by its ability to record both topographical and vegetational information.

5.5 Unmanned Aerial Vehicles (UAV) Technology

Drones, or UAVs, have changed a lot of fields thanks to their versatile nature. They can be used for anything from military and surveillance to farming and filmmaking to keeping an eye on the environment. Over the past few decades, UAV technology has advanced significantly, ushering in a new era of productivity and creativity.

UAVs are miniature planes that can fly without a pilot. Drones range in size from small, portable versions to enormous, complex ones employed in the military and industry. The airframe, the propulsion system, the control system, and the on-board sensors are the essential components of an unmanned aerial vehicle.

The ability for UAVs to fly independently is a major breakthrough. Drones today are able to undertake complex jobs like search and rescue missions or aerial surveys since they are fitted with GPS and sophisticated software. Thanks to advancements in automation, UAVs are now usable in many other fields.

When it comes to using UAVs for spying purposes, the military was an early adopter. Intelligence gathering, target identification, and even precision strikes have all benefited greatly from the use of drones in recent years. They have saved lives and decreased collateral damage thanks to their capacity to operate in high-risk areas and give real-time data.

UAVs are being used in commercial agriculture for crop monitoring, soil analysis, and watering efficiency. They are also utilised in the film industry to capture magnificent aerial pictures, the building industry to keep an eye on progress, and the environmental research sector to keep tabs on wildlife and measure ecological shifts.

Regulation and safety for UAVs are an important part of the technology. To protect civilian airspace and individual privacy, governments around the world have enacted regulations and guidelines for drone usage. Manufacturers are also working to make drones safer by developing features like collision avoidance and fail-safes (Van Genderen 2011; Varotsos and Krapivin 2020).

Improvements in unmanned aerial vehicle technology bode well for further disruptive innovation across several sectors. Drones are expected to play an increasingly important role in the future as scientists work to develop more efficient propulsion systems, extend flight periods, and advance sensor technology. UAVs are going to change the world in profound ways, whether it's through environmental monitoring, package delivery, or increased public safety. Benefits of using UAVs in wetland research are also diagrammatically represented in Fig. 5.5 and some examples in Table 5.9 for wetlands and in Table 5.10 for saline wetlands.

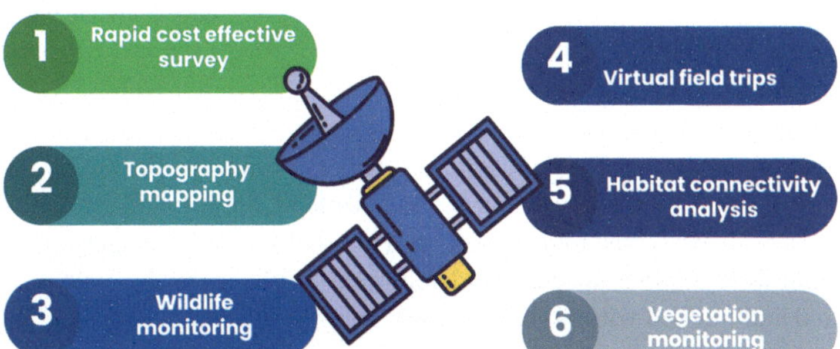

Fig. 5.5 Use of UAV for wetland studies

Table 5.9 Use of UAV for wetland studies

S. no.	Name of wetland	Type of wetland	UAV satellite used	Purpose of use
1	Everglades	Marshland	DJI Phantom 4 Pro	Monitoring wetland vegetation and water
2	Okavango Delta	Inland Delta	Parrot Anafi USA	Mapping deltaic topography and wildlife
3	Pantanal	Seasonal Floodplain	SenseFly eBee XRT	Assessing floodplain dynamics and habitat
4	Sundarbans	Mangrove Forest	Yuneec H520	Mapping mangrove cover and tiger habitats
5	Danube Delta	River Delta	DJI Matrice 300 RTK	Studying deltaic land changes and birds
6	Kakadu National Park	Wetland	PrecisionHawk Lancaster	Monitoring wetland health and crocodiles
7	Hula Valley	Marshland	Delair UX11	Water level monitoring and bird counts
8	Tonle Sap Lake	Flooded Forest	Sensefly Ebee Classic	Mapping lake conditions and fisheries
9	Lake Chad	Endorheic Lake	Parrot Disco-Pro AG	Monitoring lake area and fish migration
10	Okefenokee Swamp	Swamp	senseFly eBee XRT	Assessing swamp vegetation and alligators
11	Liaohe Delta	River Delta	Delair UX11	Deltaic landform mapping and water levels
12	Narew National Park	Wetland	DJI Matrice 300 RTK	Monitoring wetland water levels and birds
13	Tisza River Basin	River Basin	senseFly eBee XRT	Studying river basin dynamics and habitat
14	Morava River Floodplain	Floodplain	Parrot Anafi USA	Floodplain assessment and fish counts
15	Esteros del Ibera	Wetland	Yuneec H520	Mapping wetland features and capybaras
16	Poyang Lake	Seasonal Lake	Delair UX11	Monitoring lake water levels and fisheries
17	Rann of Kutch	Seasonal Salt Marsh	senseFly eBee Classic	Salt marsh dynamics and bird surveys
18	Itaipu Dam Reservoir	Reservoir	DJI Phantom 4 RTK	Reservoir water level monitoring and fish
19	Humber Estuary	Estuary	Parrot Anafi USA	Mapping estuary conditions and bird counts
20	Salton Sea	Endorheic Lake	senseFly eBee XRT	Lake level assessment and bird population

High-Resolution Imagery: High-resolution cameras mounted on UAVs allow for precise mapping of wetland ecosystems. Wetland flora, landforms, and water bodies can all be better understood with the help of this imagery. Vegetation communities, invasive species, and temporal shifts in land cover can all be detected and tracked with high-resolution images of wetland areas.

Table 5.10 Use of UAV for saline wetland studies

S. no.	Name of saline wetland	Type of saline wetland	UAV satellite used	Purpose of use
1	Chott el Jerid	Salt Flat	DJI Phantom 4 Pro	Monitoring salt flat surface and minerals
2	Sambhar Lake	Salt Lake	Parrot Anafi USA	Mapping lake water quality and sediments
3	Great Salt Lake	Terminal Lake	senseFly eBee XRT	Assessing lake level and water constituents
4	Lake Urmia	Endorheic Lake	Yuneec H520	Monitoring lake area and salt content
5	Maranjab Salt Desert	Salt Desert	Delair UX11	Mapping salt desert features and minerals
6	Etosha Pan	Salt Pan	senseFly eBee Classic	Studying pan conditions and wildlife
7	Lake Natron	Alkaline Lake	DJI Matrice 300 RTK	Mapping lake chemistry and shoreline
8	Salar de Uyuni	Salt Flat	Parrot Disco-Pro AG	Assessing salt flat topography and changes
9	Lake Eyre	Salt Pan	senseFly eBee XRT	Topographic mapping and flood assessment
10	Lop Nur	Salt Pan	Delair UX11	Monitoring salt pan conditions and topography
11	Badwater Basin	Salt Pan	Parrot Anafi USA	Assessing salt pan topography and changes
12	Lake Abert	Endorheic Lake	DJI Matrice 300 RTK	Monitoring lake water quality and sediments
13	Lake Tuz	Salt Lake	senseFly eBee Classic	Mapping lake chemistry and conditions
14	Koyashskoye Salt Lake	Salt Lake	Parrot Disco-Pro AG	Assessing salt lake chemistry and minerals
15	Sebkha Sejoumi	Salt Pan	Yuneec H520	Studying salt pan features and composition
16	Lake Tengiz	Salt Lake	senseFly eBee XRT	Monitoring lake water chemistry and sediments
17	Lake Qarun	Salt Lake	Parrot Anafi USA	Mapping lake conditions and shoreline
18	Kara-Bogaz-Gol	Lagoon	Delair UX11	Assessing lagoon water quality and features
19	Soda Lakes of Kenya	Saline Lakes	Yuneec H520	Studying saline lake chemistry and sediments
20	Lake Assal	Crater Lake	DJI Matrice 300 RTK	Mapping crater lake chemistry and sediments

Remote Sensing Capabilities: Multispectral, hyperspectral, and thermal cameras are just a few examples of remote sensing instruments that can be mounted to UAVs. In order to evaluate wetland health, vegetation stress, or water quality, these sensors capture data beyond what can be seen in visible photography. Using UAVs for remote sensing is a practical and economical method of collecting data at high spatial and temporal resolutions.

5.5.1 Topographic Mapping

DEMs of wetland areas can be generated with great accuracy by UAVs using photogrammetric techniques. UAVs can construct accurate 3D models of the wetland's topography by gathering overlapping airborne photographs. These DEMs contribute to wetland management and restoration by allowing for a more thorough examination of hydrology, water flow patterns, and elevation changes.

Collecting data from the air with UAVs allows for the creation of DEMs or topographic maps that are both precise and detailed. Drones with cameras or LiDAR sensors are used to collect data from above, and this information is used to create detailed topographic maps. Here's an overview of the process:

Flight Planning: The first thing to do is design a flight plan for the UAV. This includes deciding where to do the research, figuring out the optimal flight altitude and overlap, and taking into account any local limitations. Optimising flight paths and guaranteeing sufficient coverage can be aided by using flight planning software or specialist UAV mission planning tools.

Data Acquisition: The UAV is sent into the air to take pictures or collect LiDAR data of the area under investigation. At regular intervals throughout the flight route, the UAV takes pictures or sends out laser pulses to collect LiDAR data. For precise spatial placement, the imagery or LiDAR data is georeferenced using in-air GPS or ground control points (GCPs) on the ground.

Data Processing: A computer model of the landscape is constructed using the processed images or LiDAR data that was collected. A dense point cloud, representing the 3D structure of the area, is created by the processing of the photos for photogrammetry-based mapping using specialist software. A point cloud can be generated straight from LiDAR data. A high-resolution DEM (digital surface model) is generated from the point cloud data after it has been cleaned, categorised, and interpolated DSM.

Data Analysis and Editing: Analysis and adjustments are made to the resulting DEM or DSM. In order to achieve this goal, the elevation data may be refined, systemic mistakes may be fixed, or outliers may be removed. To guarantee the quality of the topographic data, other editing methods like manual editing or filtering can be implemented.

Map Generation: Finally, topographic maps will be created. Elevation contour lines at regular intervals can be generated from the DEM or DSM. Aerial images and other data sources can be used to enhance the map by adding additional topographic features like bodies of water, roads, and buildings. The finished map can be made in a number of different formats, including digital (such as GeoTIFF) and printable (such as PDF).

Topographical mapping with UAVs has various benefits over more conventional approaches. It offers high-resolution information for producing precise topographic maps.

5.5.2 Wetland Change Detection

UAVs can track and record changes in wetland ecosystems over time. Researchers can see changes in wetland size, vegetation cover, and water bodies by comparing aerial photos or DEMs collected at different times. This data is useful for determining the state of wetlands, measuring the results of management initiatives, and pinpointing problem areas in need of protection or restoration.

Monitoring and assessing long-term changes in wetland ecosystems is made possible through the use of UAVs. Drones may take high-resolution photos and gather other data, allowing scientists and researchers to examine wetland conditions, spot changes, and evaluate the ecosystem's well-being. Here's an overview of the process:

Planning and Flight Preparation

Define the study area: Determine the wetland area of interest that you want to monitor.

Flight planning: Determine the flight parameters, such as altitude, overlap, and flight path, to ensure comprehensive coverage and high-quality data acquisition.

Obtain necessary permissions: Check local regulations and obtain any required permits for UAV flights.

UAV Data Acquisition

Pre-flight checks: Ensure the UAV is in good working condition, including batteries, cameras/sensors, and GPS.

Flight execution: Fly the UAV over the wetland area, following the predetermined flight plan. The UAV should capture high-resolution imagery using onboard cameras or other remote sensing sensors.

Data collection: Collect additional data, such as LiDAR or multispectral imagery, if available and applicable to your study objectives.

Data Processing

Image stitching: Combine the individual images captured during the UAV flight to create a single, seamless Ortho mosaic image of the wetland.

Georeferencing: Assign geographic coordinates to the Ortho mosaic image using ground control points or GPS data.

Classification: Analyse the imagery to identify different wetland features, such as water bodies, vegetation, and land cover.

Change detection: Identify and quantify temporal shifts in wetland properties by comparing newly collected data with historical records. Both human visual interpretation and machine systems can accomplish this.

Data Analysis and Interpretation

Vegetation analysis: Determine the density of vegetation and the types of plants it contains using remote sensing.

Water quality assessment: Measure water properties like turbidity, chlorophyll concentration, and water depth with the help of multispectral or hyperspectral images.
Wetland mapping: Make fresh maps and geographical datasets detailing wetlands and their attributes.
Trend analysis: Assess wetland health and potential hazards by analysing long-term changes in wetland extent, vegetation cover, or other factors.

Reporting and Communication
Summarise findings: Provide an overview of the wetland change analysis's findings, highlighting any significant trends or consequences.
Visualisation: Effectively convey the findings through the use of maps, graphs, and other visual aids.
Recommendations: Make management, conservation, or restoration suggestions for the wetland based on the findings.

It's important to remember that ground-based observations and other data sources can improve the accuracy and interpretation of wetland change detection utilising UAVs. Furthermore, the study's goals, resources, and the type of wetland ecosystem being studied can all influence the precise methodology and tools employed (Warren and Marz 2015; Weitkamp 2006).

5.5.3 Wildlife Monitoring

Wetland animals can be monitored with the help of UAVs equipped with thermal cameras or LiDAR. Wildlife, such as birds, mammals, or reptiles, can be detected and tracked with the use of these sensors, revealing important information on the animals' habits, population trends, and preferred environments. UAVs provide a low-impact and non-intrusive method of animal monitoring in wetlands.

The capacity of UAVs to provide aerial perspectives, cover broad regions, and capture high-resolution imagery has led to their widespread adoption for wildlife monitoring. In wildlife monitoring, UAVs can be used for a wide range of tasks, such as population surveys, behaviour analysis, habitat mapping, and anti-poaching. Using UAVs for wildlife research: a brief overview:

Planning and Preparation
Define the study objectives: Determine the specific goals of the wildlife monitoring project, such as population assessment, tracking migration patterns, or studying behaviour.
Study area selection: Identify the area or specific habitats where the wildlife monitoring will take place.
Flight planning: Determine the flight parameters, including altitude, flight path, and timing, based on the study objectives and local regulations.

UAV Data Acquisition
UAV selection: Choose an appropriate UAV that meets the requirements of the study, such as flight endurance, payload capacity, and manoeuvrability.

Sensor selection: Determine the sensors or cameras needed to capture relevant data. This can include high-resolution optical cameras, thermal cameras for nocturnal monitoring, multispectral or hyperspectral sensors for vegetation analysis, or LiDAR for 3D mapping.

Flight execution: Conduct UAV flights over the study area, following the predefined flight plan. Collect imagery or other data using the onboard sensors.

Data collection: Capture images or other data at regular intervals, ensuring appropriate overlap for image stitching and analysis.

Data Processing and Analysis

Image processing: Stitch the individual images captured during the UAV flights to create a seamless orthomosaic image or 3D model of the study area.

Object detection and tracking: Analyse the imagery to identify and track wildlife species of interest. This can be done manually or using automated algorithms for object detection and machine learning techniques for classification.

Population estimation: Use statistical methods and mark-recapture techniques to estimate wildlife populations based on the identified individuals or groups.

Behavioural analysis: Study animal behaviour by analysing movement patterns, social interactions, or habitat preferences observed in the UAV imagery.

Habitat mapping: Generate detailed maps of the study area, including vegetation cover, water bodies, and other relevant habitat characteristics.

Reporting and Conservation Efforts

Data interpretation: Interpret the findings to gain insights into wildlife populations, behaviour, or habitat conditions.

Conservation planning: Use the collected data to inform conservation strategies, such as identifying key habitats, migration corridors, or areas requiring protection.

Anti-poaching efforts: Utilise UAVs for surveillance and rapid response in anti-poaching activities, such as detecting illegal hunting or poaching activities.

Stakeholder engagement: Share the results with relevant stakeholders, including conservation organisations, wildlife management authorities, and local communities, to raise awareness and support conservation efforts.

Keep in mind that UAV wildlife monitoring must adhere to all applicable laws, wildlife ethics, and privacy policies. For UAV-based wildlife monitoring to be effective and ethical, it must begin with thorough preparation, engagement with local authorities, and input from wildlife experts.

5.5.4 Rapid and Cost-Effective Surveys

UAVs provide a quick and cheap way to collect data over huge wetlands. UAVs can survey larger regions in less time and at lower prices than traditional field surveys or human aerial surveys. Because of their convenience, UAVs are increasingly being considered for use in complete wetland evaluations, even in areas that are difficult to reach.

Many industries can benefit greatly from the use of unmanned aerial vehicles (UAVs) because of the speed and low cost with which they can undertake surveys. Here are some key benefits of using UAVs for such surveys:

Time Efficiency: Unmanned aerial vehicles have the ability to survey big areas quickly. Quickly capturing aerial pictures or other data allows for faster surveys than with more conventional ground-based methods. This is especially helpful for time-sensitive tasks or those that require frequent monitoring.

Cost Savings: Comparatively, UAV surveys can save money compared to those conducted by human aircraft or on the ground. UAVs are accessible to a wide range of organisations and initiatives due to their low initial investment and ongoing operating costs. Savings in manpower, logistics, and fuel enable more efficient coverage of larger areas at lower total costs.

High-Resolution Data: UAVs of today have high-resolution imaging sensors that allow them to take images and films from above. This allows for more precise mapping, measurements, and data analysis, leading to more trustworthy survey results.

Flexibility and Accessibility: UAVs can go where humans or conventional vehicles would have a hard time going. This makes them perfect for surveys in remote or inaccessible areas, such as mountainous regions or heavily wooded regions. UAVs have the ability to fly at low altitudes, allowing for a more intimate and comprehensive inspection.

Real-Time Data Collection: When UAVs have the right sensors, they can send information back to the ground control centre in near real time. This allows for instant evaluation and decision-making, which improves the effectiveness of survey activities. In addition, problems or changes that may require immediate attention can be spotted much faster using real-time data.

Safety and Reduced Environmental Impact: In risky situations or areas with dangerous species, UAV surveys can eliminate or greatly reduce danger to human surveyors. In addition, UAVs have a lower carbon footprint than conventional survey techniques.

Integration with GIS and Remote Sensing: Survey information collected by UAVs can be easily combined with data collected by GIS programmes and remote sensing tools. Decision-makers and planners can benefit from this combination since it streamlines the processes of spatial analysis, mapping, and data display.

Monitoring and Change Detection: By collecting data at various intervals, UAVs make it possible to conduct consistent monitoring and spot any changes over time. This is helpful for evaluating ongoing activities such as landscape evolution, infrastructure construction, plant expansion, and so on. Comparing UAV data over time can provide valuable insights into trends and patterns.

Although UAVs provide many advantages, there are also certain drawbacks that should be taken into account. Some of these factors include the need for qualified operators and analysts, the need of adhering to regulations, the impact of weather on flight operations, the limits of flight duration and payload capacity, and the necessity of processing large amounts of data. Despite these limitations, UAV surveys are becoming increasingly popular in fields as diverse as agriculture, environmental monitoring, infrastructure inspection, and disaster response because of their speed and low cost.

5.5.5 Environmental Education and Outreach

UAVs can be used to obtain stunning aerial footage of wetlands, highlighting the region's ecological value. These images can be used in the classroom to teach students about the importance of wetlands and inspire them to become involved in protecting these ecosystems. UAVs, or unmanned aerial vehicles, can be extremely useful for environmental education and outreach because of the fresh viewpoints and exciting opportunities they give students. Here are some ways UAVs can be used for environmental education:

Aerial Tours and Demonstrations: UAVs can be used to give people a bird's-eye view of the area being explored or demonstrated. This is a great way to show out protected locations or ecosystems that are otherwise inaccessible to the general public. Participants can experience the natural world in all its glory, learning to value and respect it at a deeper level as a result.

Virtual Field Trips: Virtual field trips can be made that are just as realistic as the real thing by using UAVs with high-definition cameras or 360-degree video capture. Learning environments such as protected habitats or fragile ecosystems can be safely explored by students. Sharing virtual field excursions on classroom computers, social media, or VR headsets increases exposure and participation.

Environmental Surveys and Data Collection: Involve students in collecting data and doing environmental surveys using UAVs to encourage active learning. UAV flight planning, operation, and capture of aerial photography and data collection are all skills they can acquire. The significance of data collecting in environmental monitoring and conservation is driven home to students through this hands-on exercise.

Habitat Monitoring and Mapping: As part of environmental awareness campaigns, you may use unmanned aerial vehicles to survey and document natural areas. Studying aerial imagery or 3D models generated from UAV data can teach students about ecosystems, biodiversity, and ecological processes from a bird's eye view. A greater awareness of environmental dynamics is fostered by their ability to recognise habitats, evaluate their status, and track changes through time.

Citizen Science Projects: Engage kids in UAV-based citizen science initiatives. Students can help with data collection by mapping invasive species, monitoring wildlife populations, and observing land use changes, among other things. Students develop a sense of environmental responsibility and the ability to make meaningful contributions to environmental research through their involvement.

Environmental Storytelling: Captivating aerial footage shot by UAVs can be used in a variety of ways, including in presentations, documentaries, and instructional videos. Stories about environmental problems, conservation successes, or the toll that human activities take on ecosystems can all benefit from the inclusion of such images. Inspiring narratives can help people feel more connected to one another and environmental issues.

Collaborative Projects and Competitions: Motivate your pupils to team together on environmental UAV projects. They can work together in groups to plan and carry out initiatives aimed at ecological observation, habitat restoration, and public participation. Motivating students and stimulating innovation can be achieved by hosting competitions or challenges focused on UAVs and environmental concerns.

Community Engagement and Outreach: Get your pupils to team up on UAV-related environmental projects. They can work together in groups to develop and carry out initiatives for environmental monitoring, habitat restoration, and community outreach. Students might be inspired to innovate and compete through the organisation of competitions or challenges centred on UAVs and environmental concerns (Woodhouse 2005; Zeng et al. 2020; Zikopoulos and Eaton 2011).

Prioritising safety, adhering to local regulations, and encouraging responsible UAV use are crucial for using UAVs in environmental education and outreach projects. It is also important to provide educators and students with proper training and guidance on UAV operation, ethical considerations, and data protection.

Researchers and wetland managers are able to collect high-resolution data, monitor changes, evaluate habitat conditions, and make educated decisions for wetland conservation and restoration thanks to UAVs. However, UAVs must be used properly, in accordance with local legislation and guidelines, to protect the environment and other users of the airspace.

References

Arst H, Arst H (2003) Optical properties and remote sensing of multicomponental water bodies, vol 231. Springer, Chichester

Avtar R, Watanabe T (eds) (2020) Unmanned aerial vehicle: applications in agriculture and environment. Springer International Publishing, Berlin

Beard RW, McLain TW (2012) Small unmanned aircraft: theory and practice. Princeton University Press, Princeton

Chang CI (ed) (2007) Hyperspectral data exploitation: theory and applications. John Wiley & Sons, New York

Dong P, Chen Q (2017) LiDAR remote sensing and applications. CRC Press, Boca Raton

Elachi C, Van Zyl JJ (2021) Introduction to the physics and techniques of remote sensing. John Wiley & Sons, New York

Fahlstrom PG, Gleason TJ, Sadraey MH (2022) Introduction to UAV systems. John Wiley & Sons, New York

Liang S, Wang J (eds) (2019) Advanced remote sensing: terrestrial information extraction and applications. Academic Press, New York

Marghany M (ed) (2016) Environmental applications of remote sensing. BoD–Books on Demand, Norderstedt

Megie G (1985) Laser remote sensing: fundamentals and applications. Eos Trans Am Geophys Union 66:686

Miller RL, Del Castillo CE, McKee BA (eds) (2005) Remote sensing of coastal aquatic environments, vol 511. Springer, Dordrecht
Mishra DR, Ogashawara I, Gitelson AA (eds) (2017) Bio-optical modeling and remote sensing of inland waters. Elsevier, Amsterdam
Ollero A, Maza I (eds) (2007) Multiple heterogeneous unmanned aerial vehicles, vol 37. Springer, Berlin
Pandey PC, Balzter H, Srivastava PK, Petropoulos GP, Bhattacharya B (2020a) Future perspectives and challenges in hyperspectral remote sensing. In: Earth observation. Elsevier, Amsterdam, pp 429–439
Pandey PC, Srivastava PK, Balzter H, Bhattacharya B, Petropoulos GP (eds) (2020b) Hyperspectral remote sensing: theory and applications. Elsevier, Amsterdam
Pu R (2017) Hyperspectral remote sensing: fundamentals and practices. CRC Press, Boca Raton
Rees G (1999) The remote sensing data book. Cambridge University Press, Cambridge
Richards JA (2009) Remote sensing with imaging radar, vol 1. Springer, Berlin, pp 172–173
Sabins FF Jr, Ellis JM (2020) Remote sensing: principles, interpretation, and applications. Waveland Press, Long Grove
Shokr M, Sinha NK (2023) Sea ice: physics and remote sensing. John Wiley & Sons, New York
Thenkabail PS, Lyon JG (eds) (2016) Hyperspectral remote sensing of vegetation. CRC Press, Boca Raton
Toro FG, Tsourdos A (eds) (2018) UAV sensors for environmental monitoring. MDPI, Basel
Van Genderen JL (2011) Advances in environmental remote sensing: sensors, algorithms, and applications, vol 4. Routledge, London, p 446
Varotsos CA, Krapivin VF (2020) Microwave remote sensing tools in environmental science. Springer International Publishing, New York
Warren J, Marz N (2015) Big data: principles and best practices of scalable realtime data systems. Simon and Schuster, New York
Weitkamp C (ed) (2006) Lidar: range-resolved optical remote sensing of the atmosphere, vol 102. Springer Science & Business, Berlin
Woodhouse IH (2005) Introduction to microwave remote sensing. CRC Press, Boca Raton
Zeng Y, Guvenc I, Zhang R, Geraci G, Matolak DW (eds) (2020) UAV communications for 5G and beyond. John Wiley & Sons, New York
Zikopoulos P, Eaton C (2011) Understanding big data: analytics for enterprise class hadoop and streaming data. McGraw-Hill Osborne Media, New York

Future Technologies for Wetland Studies

6

Abstract

This chapter ventures into the cutting-edge realm of technological innovations poised to transform environmental science and conservation practices. This chapter concentrates on three emerging technologies—Artificial Intelligence (AI), Big Data, and the Internet of Things (IoT)—that hold immense potential for shaping the future of these fields. "Artificial intelligence (AI)" is introduced in the first section, highlighting its capacity to revolutionise environmental research and conservation efforts. With its machine learning algorithms and cognitive capabilities, AI can process vast datasets swiftly and accurately, facilitating tasks such as species identification, environmental trend prediction, and resource management optimisation. The discussion then shifts to "Big Data" in the second section, emphasising its role in handling massive datasets collected from various sources, including remote sensors and citizen science projects. Big Data analytics unveil hidden patterns and trends, enabling real-time monitoring of environmental changes and informing decisions for disaster management and climate change mitigation. Lastly, the exploration extends to the "Internet of Things (IoT)" in the third section, which entails a network of interconnected devices and sensors gathering and sharing environmental data. IoT technology enables continuous monitoring of wildlife, water quality, and weather conditions, empowering swift responses to environmental threats and fostering a global network for environmental conservation. Through these emerging technologies, this chapter illuminates a promising future for environmental science and conservation, driven by innovation and data-driven decision-making.

Keywords

Artificial Intelligence (AI) · Big Data · Internet of Things (IoT)

L. K. Sharma, R. Naik, *Conservation of Saline Wetland Ecosystems*,
https://doi.org/10.1007/978-981-97-5069-6_6

6.1 Artificial Intelligence (AI)

AI, or artificial intelligence, is a cutting-edge subfield of computer science concerned with the development of self-learning, autonomous machines that can do complex human-like activities. Recent years have seen exponential growth in AI, which is reshaping many facets of human existence from healthcare and banking to transportation and entertainment.

At its foundation, AI strives to construct computer systems that can think, learn, and adapt in ways that replicate human cognitive functions. Machine Learning (ML), Natural Language Processing (NLP), Computer Vision (CV), and robotics are just a few of the many subfields that fall under its umbrella. Machine learning is a subfield of AI concerned with teaching computers to learn from data and make judgments or predictions accordingly.

One of the most impressive AI developments is deep learning, a kind of machine learning that employs multi-layered neural networks to sift through and draw conclusions from enormous data sets. This innovation has fuelled the development of various AI applications like speech and picture recognition, driverless cars, and personalised recommendations. It has also enabled the development of chatbots and virtual assistants, like Siri and Alexa, that can interpret and respond to natural language.

The use of AI is changing many sectors. It's a big help in the medical field for things like diagnosis, treatment planning, and prognosis. AI algorithms are used in the financial sector to evaluate market data, manage investments, and spot fraudulent transactions. The advent of autonomous cars holds great promise for improving the convenience and security of transportation. AI is also enhancing customer experiences with chatbots and personalised recommendations, and it is transforming supply chain management and manufacturing.

But AI also brings up moral and social issues that need to be addressed. Concerns about privacy, loss of employment owing to automation, and the introduction of bias into AI decision-making are all examples. These issues, as well as the creation of legislation and norms, must be given serious thought before AI systems can be developed and released responsibly.

AI is a game-changing tool with the ability to alter virtually every area of human existence. The speed with which it has evolved and been adopted by many sectors demonstrates its value as a driver of innovation. While AI's potential is undeniable, the field still faces significant ethical and social obstacles that must be overcome before it can be used to truly advance humanity. Here are some key applications of AI in wetland studies which are also diagrammatically represented in Fig. 6.1 and some examples in Table 6.1 for wetlands and in Table 6.2 for saline wetlands.

Wetland Mapping and Classification: Remote sensing data, such as satellite images or aerial photography, can be automatically categorised and mapped to wetland areas through the use of AI algorithms. Wetland features including water bodies, vegetation kinds, and land cover classifications can be accurately identified by AI

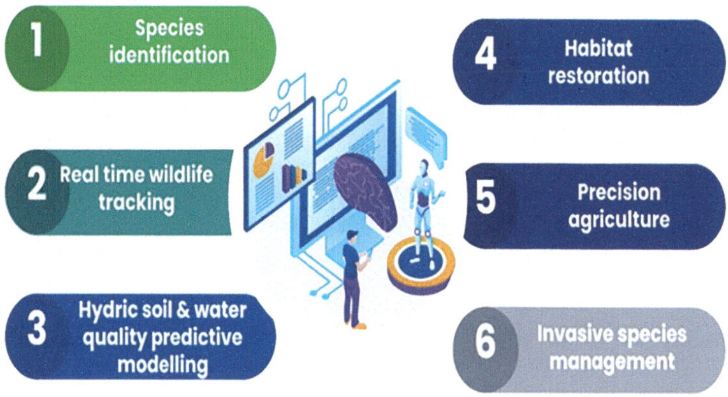

Fig. 6.1 Use of AI for wetland studies

models trained on labelled training data. The process of inventorying, monitoring, and managing wetlands is aided by this (Barrat 2023).

6.1.1 Wetland Mapping and Classification

In order to automatically detect and categorise wetland areas, remote sensing data is analysed with AI algorithms for wetland mapping and classification. The steps involved are as follows:

Data Acquisition: Acquire satellite images, aerial photos, or LiDAR scans of the wetland region in question. For precise categorisation, the data must have sufficient spatial resolution and spectral information.

Data Preprocessing: The quality of the remote sensing data used in the AI analysis can be improved with some preliminary processing. Image calibration, atmospheric correction, and radiometric and geometric measurement adjustments may be necessary.

Training Data Collection: Get your hands on a large amount of properly labelled training data. Here, we pick out locations in the remote sensing data that stand in for several types of wetlands (e.g., open water, emergent vegetation, submerged vegetation, upland). Wetland-type labels should be placed on each sample.

Feature Extraction: Use the remote sensing data to extract useful spectral, textural, and spatial properties. These attributes capture the unique qualities of various wetland types and are used as training data for AI algorithms.

AI Model Training: Using the labelled training data and the extracted features, train an artificial intelligence model, often a machine learning technique like convolutional neural networks (CNNs) or random forest (RF). The model is trained to recognise the associations between the feature types and the corresponding wetland types.

Table 6.1 Use of AI for wetland studies

S. no.	Name of wetland	Type of wetland	AI technology used	Purpose of use
1	Everglades	Marshland	AI Image Analysis	Monitoring wetland vegetation changes
2	Okavango Delta	Inland Delta	Machine Learning Models	Mapping and predicting deltaic land changes
3	Pantanal	Seasonal Floodplain	AI Remote Sensing	Assessing flood dynamics and wildlife habitat
4	Sundarbans	Mangrove Forest	AI Object Recognition	Mapping mangrove cover and tiger habitats
5	Danube Delta	River Delta	Deep Learning Algorithms	Monitoring deltaic landform changes
6	Kakadu National Park	Wetland	AI Data Fusion	Studying wetland ecology and species diversity
7	Hula Valley	Marshland	AI Pattern Recognition	Monitoring water levels and bird populations
8	Tonle Sap Lake	Flooded Forest	AI Spectral Analysis	Mapping lake conditions and fisheries
9	Lake Chad	Endorheic Lake	AI Predictive Modeling	Assessing lake area and water level changes
10	Okefenokee Swamp	Swamp	AI Image Classification	Studying swamp vegetation and wildlife
11	Liaohe Delta	River Delta	AI Change Detection	Monitoring deltaic landform dynamics
12	Narew National Park	Wetland	AI Environmental Modeling	Tracking wetland water levels and biodiversity
13	Tisza River Basin	River Basin	AI Hydrological Analysis	Assessing river basin dynamics and flooding
14	Morava River Floodplain	Floodplain	AI Data Integration	Analysing floodplain changes and fish populations
15	Esteros del Ibera	Wetland	AI Biodiversity Assessment	Mapping wetland features and species diversity
16	Poyang Lake	Seasonal Lake	AI Ecosystem Modeling	Monitoring lake conditions and bird populations
17	Rann of Kutch	Seasonal Salt Marsh	AI Habitat Assessment	Studying salt marsh dynamics and bird habitats
18	Itaipu Dam Reservoir	Reservoir	AI Water Quality Analysis	Monitoring reservoir water quality and fish
19	Humber Estuary	Estuary	AI Estuarine Ecology	Mapping estuary conditions and fish populations
20	Lake Tahoe	Alpine Lake	AI Water Clarity Analysis	Assessing lake water clarity and environmental
21	Salton Sea	Endorheic Lake	AI Environmental Monitoring	Monitoring lake conditions and bird populations

Model Validation and Refinement: Use independent validation data to assess the quality of the AI model's training data. Wetland classification findings should be evaluated, and any necessary adjustments to the model or more training data should be made.

Table 6.2 Use of AI for saline wetlands

S. no.	Name of saline wetland	Type of saline wetland	AI technology used	Purpose of use
1	Chott el Jerid	Salt Flat	AI Image Analysis	Monitoring salt flat surface changes
2	Sambhar Lake	Salt Lake	AI Object Detection	Mapping lake water quality and sediments
3	Great Salt Lake	Terminal Lake	AI Remote Sensing	Assessing lake level and water constituent changes
4	Lake Urmia	Endorheic Lake	AI Data Analytics	Monitoring lake area and salt content
5	Maranjab Salt Desert	Salt Desert	AI Image Classification	Mapping salt desert features and minerals
6	Etosha Pan	Salt Pan	AI Pattern Recognition	Studying pan conditions and wildlife
7	Lake Natron	Alkaline Lake	AI Spectral Analysis	Mapping lake chemistry and shoreline changes
8	Salar de Uyuni	Salt Flat	AI Change Detection	Assessing salt flat topography and changes
9	Lake Eyre	Salt Pan	AI Environmental Modeling	Topographic mapping and flood assessment
10	Lop Nur	Salt Pan	AI Hydrological Analysis	Monitoring salt pan conditions and topography
11	Badwater Basin	Salt Pan	AI Biodiversity Assessment	Assessing salt pan topography and changes
12	Lake Abert	Endorheic Lake	AI Habitat Analysis	Monitoring lake water quality and sediments
13	Lake Tuz	Salt Lake	AI Ecosystem Modeling	Mapping lake chemistry and conditions
14	Koyashskoye Salt Lake	Salt Lake	AI Water Quality Analysis	Assessing salt lake chemistry and minerals
15	Sebkha Sejoumi	Salt Pan	AI Data Fusion	Studying salt pan features and composition
16	Lake Tengiz	Salt Lake	AI Climate Modeling	Monitoring lake water chemistry and sediments
17	Lake Qarun	Salt Lake	AI Vegetation Analysis	Mapping lake conditions and shoreline changes
18	Kara-Bogaz-Gol	Lagoon	AI Wildlife Tracking	Assessing lagoon water quality and features
19	Soda Lakes of Kenya	Saline Lakes	AI Sediment Analysis	Studying saline lake chemistry and sediments
20	Lake Assal	Crater Lake	AI Satellite Imagery Analysis	Mapping crater lake chemistry and sediments

6.1.1.1 Wetland Mapping and Classification

To properly categorise the wetlands, use the AI model that was developed on the whole remote sensing dataset. The AI algorithm examines the data's spectral fingerprints and geographical patterns to mechanically categorise pixels as marsh or dry terrain.

Post-processing and Accuracy Assessment: The wetland classification map should be post-processed to get rid of any artefacts or noise. To evaluate the overall accuracy and generate an error matrix, compare the categorised map to ground truth data or field surveys.

Mapping Products and Interpretation: Create themed maps of the various wetland classes, wetland extent maps, and change detection maps based on the findings of the categorisation. Learn more about wetland distribution, composition, and changes through time by deciphering the mapping data.

The effectiveness of AI-based wetland mapping and classification hinges on a number of aspects, including the availability and quality of training data, the AI algorithms employed, the features used, and the precision of ground truth data used for validation. Results from AI-based wetland mapping can be made more trustworthy and accurate with the help of human specialists and validation via field surveys or other data sources (Chaudhuri 2018).

6.1.1.2 Change Detection and Monitoring

Changes in wetland ecosystems can be detected and quantified using artificial intelligence algorithms. Artificial intelligence models can detect shifts in wetland distribution, vegetation cover, water quality, and land use by comparing past and present data. Wetland health may be evaluated, hazards can be identified, and conservation or restoration measures can be planned with the use of this data.

Using AI algorithms to examine remote sensing data, change detection, and monitoring can automatically identify and quantify changes over time in a region of interest. An outline of the steps is provided below:

Data Acquisition: Collect remote sensing information covering the area of interest at various times, such as satellite images or aerial photography. There should be enough detail in both space and time in the data to detect the alterations of interest.

Data Preprocessing: Improve the quality and applicability of the remote sensing data before using it for a change detection study. Data normalisation, co-registration of photos from different times, and radiometric and geometric corrections are all examples of what might fall under this category.

Feature Extraction: To better capture the changes of interest, relevant features should be extracted from the remote sensing data. Features such as spectral indices, textural measurements, and spatial patterns can be used to draw attention to the transformations taking place in the terrain.

AI Model Training: Apply the extracted features to the labelled training data to train an artificial intelligence model, often a machine learning method like a convolutional neural network or a recurrent neural network. The model is trained to recognise the connections and patterns between the landscape's features and its dynamic state changes.

Model Validation and Refinement: Use independent validation data to assess the quality of the AI model's training data. Verify how well the model performs at detecting changes, and if necessary, make any adjustments to the model parameters or gather more training data.

Change Detection and Mapping: Use the AI model you've trained on the whole of your remote sensing data to spot and chart out any shifts. To automatically detect and classify the changes, such as land cover conversions, urban expansion, vegetation dynamics, or water body variations, the AI programme examines the spectral signatures, texture, or spatial patterns in the data.

Post-processing and Accuracy Assessment: Clean up the findings of the change detection thereafter to get rid of any artefacts or noise. The total accuracy can be determined by comparing the observed changes to ground truth data or reference data, and an error matrix can be generated from this comparison.

6.1.1.3 Change Analysis and Interpretation

Learn more about the origin, scale, and direction of the changes you've spotted by analysing the data from your change detection efforts. Apply the findings to the study's intended purposes, such as evaluating urban growth, keeping tabs on deforestation, or tracing changes in wetland ecosystems.

Monitoring and Reporting: To ensure constant monitoring, it is necessary to repeatedly run the analysis to detect changes using newly collected remote sensing data. Produce monitoring reports or visualisations to inform stakeholders, decision-makers, or the general public of the changes, trends, and consequences that were uncovered.

It's important to remember that the success of AI-based change detection and monitoring depends on a number of different factors, including the quality and availability of training data, the AI algorithms used, the appropriateness of the selected features, and the precision of ground truth or reference data used for validation. The trustworthiness and precision of AI-based change detection can be improved through collaboration with topic experts, validation via field surveys, or integration with other data sources.

6.1.2 Species Identification and Monitoring

In wetland ecosystems, AI can automate the process of identifying species and keeping tabs on them. Wetland-dependent species, such as birds, amphibians, or plant species, can be identified and followed using image recognition techniques used in AI models. This facilitates effective species surveys and population monitoring, which contributes to efforts to preserve biodiversity.

AI algorithms can be used to automatically identify and monitor different species based on a wide variety of data, including photos, noises, and sensor readings. Here's an overview of the process:

Data Collection: Collect a set of data with instances of the target species that have been labelled. Gathering photos, audio recordings, and other useful information from sources like field surveys, wildlife databases, and online repositories is one way to accomplish this.

Data Preparation: Prepare the data for artificial intelligence analysis. As part of this process, it may be necessary to create uniform data formats, clean the data of any noise or artefacts, and arrange the information in a logical fashion.

Feature Extraction: Get the features out of the data that really capture what's important about the species. Features in photos can be anything from the overall composition to particular motifs in the details. Spectrograms, acoustic qualities, and distinctive vocalisations are all examples of features that can be extracted from audio data. The AI algorithms will take these characteristics as inputs.

AI Model Training: Labelled training data and extracted features are used to train an AI model, such as a Convolutional Neural Network (CNN) for image data or a Recurrent Neural Network (RNN) for sequential data such as audio. The model is taught to identify the unique characteristics of each species.

Model Validation and Refinement: Use independent validation data to assess the quality of the AI model's training data. If the model has to be improved, you can do so by modifying the model's parameters, gathering more training data, or enhancing the feature extraction procedure.

Species Identification: To automatically determine the species present in unlabelled fresh data, use the AI model that has been trained on labelled data. The data is analysed by the AI algorithm, and the learnt patterns and attributes are compared to the data to determine the species.

Monitoring and Tracking: Make use of the learned AI model to keep tabs on the distribution and number of a certain species over time. In order to recognise and classify species in new observations, it is necessary to continuously gather new data and apply the model. This can aid in determining population trends, migration routes, or habitat preferences.

Validation and Field Verification: Verify the accuracy of the identifications made by AI with the help of field surveys or human experts. This verifies the precision and consistency of the AI model's forecasts and aids in the detection of anomalies and misclassifications.

Reporting and Analysis: Use the data collected from species identification and monitoring to learn more about ecological or demographic trends. Create written or visual reports to share your findings with interested parties, scholars, and conservation groups.

The success of AI-based species identification and monitoring relies on elements including the quality and diversity of training data, the selection of AI algorithms appropriate for the data type, and the precision of ground truth or expert validation. Results from AI-based species identification and monitoring can be made more trustworthy through the use of methods such as domain expert collaboration, continual model development, and data integration (De Brouwer 2020).

6.1.3 Water Quality Assessment

Wetland sites generate data on water quality, such as temperature, pH, dissolved oxygen, and nutrient levels, which can be analysed by AI algorithms. Artificial intelligence algorithms can forecast and evaluate wetland water quality by linking these characteristics with environmental factors. The effects of pollutants and climate change on wetland health can be better comprehended with this data.

The use of AI algorithms for analysing water quality data and making predictions or classifications about the state of water bodies is known as "water quality assessment using AI". Here's an overview of the process:

Data Collection: Collect information about water quality from a variety of sources, such as in situ sensors, laboratory measurements, and remote sensing. Temperature, pH, dissolved oxygen, turbidity, nutrient concentrations, and pollution levels are only some of the factors that should be measured and recorded.

Data Preprocessing: Remove any outliers, deal with any missing data, and perform any necessary normalisation on the water quality data. The data is then ready for artificial intelligence analysis.

Feature Selection/Extraction: Pick out or glean important information from the water quality statistics. Key indicators of water quality can be captured by a variety of elements, such as statistical measures, temporal patterns, spatial characteristics, and derived indices.

AI Model Selection: Select a suitable AI algorithm for water quality assessment based on the data's characteristics and the task at hand. Machine learning algorithms like decision trees, random forests, SVM, and deep learning models like neural networks are common examples of AI methods in use today.

Training Data Preparation: Create a training set and a test set for water quality data. Class labels for water quality, such as "excellent water quality", "moderate pollution", or "bad water quality", should be assigned to the target variables in the training dataset using regulatory or predetermined thresholds.

AI Model Training: Using the annotated training data and the extracted features, train the chosen AI model. The water quality classes or variables are assigned to the input features, and the AI model learns the patterns and correlations between them.

Model Validation and Refinement: Check how well the AI performs by using a different dataset than the one it was trained on. Validate the model's performance by evaluating its accuracy, precision, recall, and any other metrics of interest. If necessary, tweak the model's hyperparameters or gather more training data to get the best results.

Water Quality Prediction/Classification: Use the artificial intelligence model you've trained on new, unlabelled water quality data to make predictions or assign labels to the data. The AI programme processes the data features and makes predictions or labels the data points according to their water quality.

Interpretation and Reporting: To share the findings with stakeholders, water resource managers, and regulatory bodies, you must interpret the outcomes of the AI-based water quality assessment and provide reports, visualisations, or alerts. Information on patterns, problem areas, or probable pollution sources that need to be addressed and cleaned up can all be found in such reports.

Maintaining the accuracy and dependability of AI-based water quality assessment systems requires constant monitoring, frequent model updates, and validation against ground truth data. AI can be made more useful in assessing water quality by collaboration with water quality professionals and the incorporation of AI with conventional monitoring techniques.

Wetland Hydrology Modelling: AI techniques, such as machine learning, can be employed to develop hydrological models for wetland systems. These models can simulate water flow, surface runoff, or groundwater interactions within wetlands, aiding in water resource management, flood prediction, or restoration planning. AI can help improve the accuracy and predictive capabilities of these models.

6.1.4 Wetland Hydrology Modelling

In order to model and predict hydrological processes within wetland systems, such as water flow, surface runoff, groundwater interactions, and more, AI algorithms are being increasingly used. Here's an overview of the process:

Data Collection: Collect data about the wetland's environment, including its topography, climate, soil, and hydrological parameters (such as water depth, precipitation, and evaporation). Hydrological models can't be built without this information.

Preprocessing: Clean up the data ahead of time to make sure it's usable. Steps like data cleansing, interpolation, and harmonisation may be required to compile a complete and usable dataset for AI simulations.

AI Model Selection: Based on the information at hand and the desired outcomes of the model, select a suitable AI algorithm for wetland hydrology modelling. Hydrologists frequently employ artificial intelligence methods including ANN, SVM, decision trees, and RFs.

Training Data Preparation: Separate the information you gathered into a training set and a test set. The input variables (such as precipitation, soil qualities, and topography) that affect wetland hydrological processes should be included in the training dataset alongside matching hydrological data.

AI Model Training: Use the labelled training data to teach the chosen AI model. The wetland's hydrological responses are learned by the AI model, along with the intricate correlations between the input variables. During training, the parameters of the model are adjusted so that prediction errors are reduced to a minimum.

Model Validation and Refinement: Check how well the AI performs by using a different dataset than the one it was trained on. Compare the model's forecasts to actual hydrological data and evaluate its accuracy and reliability. If the model's performance is subpar, you can tweak its hyperparameters or gather more training data to enhance it.

Hydrological Simulation and Prediction: Use the artificial intelligence model you've built to simulate and make predictions about different wetland hydrological processes. This entails modelling, on the basis of input data including precipitation, soil moisture, topographic features, water flow, runoff generation, infiltration, groundwater interactions, and other relevant variables.

Model Evaluation and Sensitivity Analysis: Compare the hydrological model's predictions to both observed data and defined hydrological metrics in order to gauge the model's efficacy. To determine which factors have the greatest impact on the model's predictions, sensitivity analysis can be run.

Interpretation and Applications: Wetland hydrological dynamics, water balance, and the impacts of various elements can be better understood by an interpretation of the modelling results. Draw on the results of the models for tasks as varied as flood prediction, ecological evaluations, and wetland restoration preparation.

Wetland hydrology modelling with AI relies on a number of elements, including access to high-quality data, adequate model training, validation against observed data, and the presence of domain experts. To effectively develop and deploy AI-based methodologies in wetland hydrology modelling, collaboration between hydrologists, wetland scientists, and AI experts is needed.

6.1.5 Wetland Ecosystem Services Assessment

Artificial intelligence can be used to evaluate the carbon sequestration, water purification, and flood mitigation that wetland ecosystems provide. Wetland services may be quantified through the application of AI models and the integration of disparate datasets, allowing policymakers and land managers to make more educated decisions about wetland conservation and management.

Water flow, surface runoff, groundwater interactions, and other hydrological processes within wetland systems can be modelled with the help of artificial intelligence (AI) algorithms. Here's an overview of the process:

Data Collection: Collect data about the wetland's environment, including its topography, climate, soil, and hydrological parameters (such as water depth, precipitation, and evaporation). Hydrological models can't be built without this information.

Data Preprocessing: Clean up the data ahead of time to make sure it's usable. Creating a complete and usable dataset for AI modelling may need cleaning, interpolating, and transforming the raw data.

Feature Selection/Extraction: Determine which aspects of the gathered information are important to wetland hydrological processes, and then choose or extract those aspects. Topographic indices, climatic factors, soil parameters, plant community traits, and land use details are all examples of such elements.

AI Model Selection: Based on the information at hand and the desired outcomes of the model, select a suitable AI algorithm for wetland hydrology modelling. Hydrologists frequently employ artificial intelligence methods including ANN, SVM, decision trees, and RFs.

Training Data Preparation: Separate the cleaned data into a training set and a test set. Wetland hydrological processes are influenced by a variety of input variables, which should be accounted for in the training dataset. The information should be applicable to a wide variety of hydrological scenarios.

AI Model Training: Use the labelled training data to teach the chosen AI model. The wetland's hydrological responses are learned by the AI model, along with the intricate correlations between the input variables. During training, the parameters of the model are adjusted so that prediction errors are reduced to a minimum.

Model Validation and Refinement: Check how well the AI performs by using a different dataset than the one it was trained on. Compare the model's forecasts to actual hydrological data and evaluate its accuracy and reliability. If the model's performance is subpar, you can tweak its hyperparameters or gather more training data to enhance it.

Hydrological Simulation and Prediction: Use the artificial intelligence model you've built to simulate and make predictions about different wetland hydrological processes. This entails modelling, on the basis of input data including precipitation, soil moisture, topographic features, water flow, runoff generation, infiltration, groundwater interactions, and other relevant variables.

Model Evaluation and Sensitivity Analysis: Compare the hydrological model's predictions to both observed data and defined hydrological metrics in order to gauge the model's efficacy. To determine which factors have the greatest impact on the model's predictions, sensitivity analysis can be run.

Interpretation and Applications: Wetland hydrological dynamics, water balance, and the impacts of various elements can be better understood by an interpretation of the modelling results. Draw on the results of the models for tasks as varied as flood prediction, ecological evaluations, and wetland restoration preparation.

Wetland hydrology modelling with AI relies on a number of elements, including access to high-quality data, adequate model training, validation against observed data, and the presence of domain experts. Wetland hydrology modelling can only be developed and applied successfully with the help of hydrologists, wetland scientists, and AI professionals working together (Greengard 2021).

6.1.6 Decision Support Systems

Wetland management decision support systems that use AI can be created. Artificial intelligence models can make suggestions for wetland restoration, land-use planning, or conservation prioritising by combining data from a variety of sources, such as ecological, hydrological, and socio-economic information. These methods improve the efficiency of wetland management techniques and allow for more evidence-based decision-making.

Integrating AI methods into DSS is what we mean when we talk about utilising AI to support decision-making. Data is analysed by these systems using AI algorithms and models, and insights and recommendations are then provided to aid with decision-making. Here's an overview of the components and processes involved in building AI-powered DSS:

Data Collection: Collect information from numerous places including databases, sensors, and other external systems. Records from the past, data collected in real time, statistics on market trends, information about customers, and anything else that could be useful in making decisions.

Data Preprocessing: Preprocess the data by cleaning, transforming, and standardising it so that it can be used effectively. Data cleaning, outlier detection, normalisation, and integration are all examples of steps that may be taken to improve the quality of data and make it more suitable for analysis.

AI Model Development: Based on the nature of the decision at hand, a suitable AI method and model can be chosen. Methods including machine learning, NLP, deep learning, and expert systems may be utilised. Create and train AI models with cleaned and organised data to uncover hidden insights and useful patterns for better decision-making.

Decision Analysis: Use the AI models you've trained to do data analysis and discover insights. Methods like classification, regression, cluster analysis, and anomaly detection can be used to spot trends, make predictions, divide up data, and reveal outliers that should be taken into account.

Decision Support: Make use of the information gleaned from the AI models to aid in making choices. Depending on the nature of the decision problem at hand, this may involve the creation of recommendations, alternative scenarios, risk assessments, or optimisation solutions. Artificial intelligence models help people make better choices by providing them with more relevant data and insights.

Visualisation and User Interface: Create intuitive interfaces and visual representations to deliver the findings of your analysis in an engaging and accessible format. Charts, graphs, dashboards, and other visual representations are examples of visualisation approaches that can be used to make AI-generated data easier to explore and understand.

Integration and Deployment: The DSS driven by AI should be incorporated into ongoing operations and procedures. This can require linking the DSS to other systems, either internally or outside. Make sure the DSS is available to and used by the people who need it by having it deployed to them.

Continuous Learning and Improvement: Keep an eye on how the DSS and AI models are doing over time. When new information is obtained or the decision-making environment changes, it is important to collect feedback, assess the models' predictive abilities, and update them accordingly.

Financial services, medical care, logistics, advertising, and other industries can all benefit from DSS powered by AI. Decision-makers can benefit from these systems because they offer data-driven insights, lessen uncertainty, and enhance the decision-making process as a whole. For AI-powered DSS to be effectively deployed and utilised, it is crucial that domain experts, data scientists, and decision-makers work together.

To reiterate, excellent data, appropriate model training, and validation are crucial to the success of AI applications in wetland studies. To effectively develop and deploy AI-based technologies in wetland research and management, ecologists, hydrologists, remote sensing experts, and data scientists must work together (Hassan and Madani 2017).

6.2 Big Data

Big Data is the collective term for the huge amounts of data, both organised and unstructured, that are created every day by businesses and individuals. The volume, velocity, and variety of this data are its defining characteristics.

Big Data describes the enormous amounts of data, both structured and unstructured, that are produced every day by businesses and individuals. The three Vs that best describe this information are its volume, velocity, and variety.

Extracting meaningful insights, patterns, and trends from these massive databases is the goal of Big Data analytics. Insights like this can improve corporate operations and forecast market movements. Big Data analytics are used by businesses across many sectors, including advertising, healthcare, banking, and retail.

Moreover, Big Data's effects are not limited to the corporate sphere. It's crucial to the advancement of science, as it facilitates the handling and analysis of massive datasets for everything from climate modelling to genetics. Public policy and urban planning are two more areas where governments utilise Big Data. Big Data has the potential to revolutionise many industries, but it also raises legitimate concerns about data privacy, security, and ethics. Responsible data handling requires attention to these concerns.

Big Data is now an indispensable aspect of our information-based society. Its volume, velocity, and variety necessitate the use of cutting-edge technologies and analytical methods with broad-reaching implications. As our dependence on data increases, the ability to successfully leverage Big Data will become increasingly important for driving breakthroughs and maintaining a competitive edge.

Some important uses for large data analysis are listed below which are also diagrammatically represented in Fig. 6.2 and some examples in Table 6.3 for wetlands and in Table 6.4 for saline wetlands.

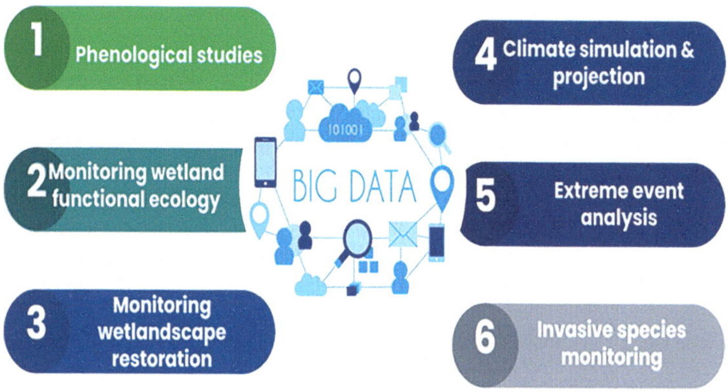

Fig. 6.2 Use of big data for wetland studies

Remote Sensing and Earth Observation: In terms of wetland features like vegetation cover, land surface temperature, water extent, and wetland shape, big data from satellite imaging, airborne sensors, and other remote sensing platforms provide detailed and regular observations. Longitudinal patterns can be examined, habitat quality evaluated, and wetland changes tracked with this data.

Wetland research relies heavily on the vast amounts of data provided by remote sensing and Earth observation, which shed light on the wetlands' properties, dynamics, and temporal changes. Here are some key applications of remote sensing and Earth observation big data in wetland studies:

6.2.1 Wetland Mapping and Classification

Satellite imaging, airborne sensors, and UAVs are all examples of remote sensing data that can be used to map and categorise wetlands. Using high-resolution images, wetland ecosystems can have their borders, vegetation kinds, water bodies, and land cover clearly defined.

6.2.2 Wetland Change Detection

Changes in wetland size, vegetation cover, land surface conditions, and water dynamics can all be evaluated with the help of multi-temporal remote sensing data. This enables successful wetland management and conservation techniques by keeping track of wetland loss, expansion, or degradation over time.

Table 6.3 Use of big data for wetland studies

S. no.	Name of wetland	Type of wetland	Big data technology used	Purpose of use
1	Everglades	Marshland	High-Performance Computing	Analysing wetland ecosystem changes
2	Okavango Delta	Inland Delta	Cloud-based Data Platform	Processing and storing deltaic data
3	Pantanal	Seasonal Floodplain	Distributed Computing	Analysing flood dynamics and wildlife habitat
4	Sundarbans	Mangrove Forest	Data Analytics Framework	Mapping mangrove cover and tiger habitats
5	Danube Delta	River Delta	Hadoop Cluster	Managing and analysing deltaic data
6	Kakadu National Park	Wetland	Spark Cluster	Studying wetland ecology and species diversity
7	Hula Valley	Marshland	Apache Cassandra	Monitoring water levels and bird populations
8	Tonle Sap Lake	Flooded Forest	Big Data Storage System	Storing and managing lake conditions data
9	Lake Chad	Endorheic Lake	Distributed Data Processing	Processing lake area and water level data
10	Okefenokee Swamp	Swamp	NoSQL Database	Analysing swamp vegetation and wildlife
11	Liaohe Delta	River Delta	Data Warehousing System	Managing and querying deltaic data
12	Narew National Park	Wetland	Data Analytics Platform	Tracking wetland water levels and biodiversity
13	Tisza River Basin	River Basin	Distributed Data Platform	Analysing river basin dynamics and flooding
14	Morava River Floodplain	Floodplain	Scalable Data Storage	Storing floodplain changes and fish data
15	Esteros del Ibera	Wetland	Data Mining Software	Mapping wetland features and species diversity
16	Poyang Lake	Seasonal Lake	Cloud Data Storage	Managing lake conditions and bird populations
17	Rann of Kutch	Seasonal Salt Marsh	Big Data Analytics Tools	Analysing salt marsh dynamics and bird habitats
18	Itaipu Dam Reservoir	Reservoir	Distributed Data Cluster	Monitoring reservoir water quality and fish
19	Humber Estuary	Estuary	Stream Processing Tools	Real-time estuary conditions and fish data
20	Lake Tahoe	Alpine Lake	Data Analytics Framework	Analysing lake water clarity and environment
21	Salton Sea	Endorheic Lake	Distributed Data Processing	Processing lake conditions and bird data

Table 6.4 Use of big data for saline wetland studies

S. no.	Name of saline wetland	Type of saline wetland	Big data technology used	Purpose of use
1	Chott el Jerid	Salt Flat	Hadoop Cluster	Analysing salt flat surface changes
2	Sambhar Lake	Salt Lake	Spark Cluster	Processing and storing lake water quality data
3	Great Salt Lake	Terminal Lake	Cloud-based Data Platform	Analysing lake level and water constituent changes
4	Lake Urmia	Endorheic Lake	Distributed Data Processing	Managing and processing lake area and salt content data
5	Maranjab Salt Desert	Salt Desert	Data Analytics Framework	Mapping salt desert features and minerals
6	Etosha Pan	Salt Pan	Apache Cassandra	Monitoring pan conditions and wildlife
7	Lake Natron	Alkaline Lake	Big Data Storage System	Storing and managing lake chemistry and shoreline data
8	Salar de Uyuni	Salt Flat	Distributed Data Processing	Analysing salt flat topography and changes
9	Lake Eyre	Salt Pan	Data Warehousing System	Managing and querying pan topography data
10	Lop Nur	Salt Pan	Data Mining Software	Analysing salt pan conditions and topography
11	Badwater Basin	Salt Pan	Cloud Data Storage	Storing and managing pan topography and changes
12	Lake Abert	Endorheic Lake	Distributed Data Cluster	Processing and analysing lake water quality and sediments
13	Lake Tuz	Salt Lake	NoSQL Database	Studying lake chemistry and conditions
14	Koyashskoye Salt Lake	Salt Lake	Scalable Data Storage	Managing salt lake chemistry and minerals
15	Sebkha Sejoumi	Salt Pan	Stream Processing Tools	Monitoring salt pan conditions and composition
16	Lake Tengiz	Salt Lake	Data Analytics Platform	Studying lake water chemistry and sediments
17	Lake Qarun	Salt Lake	Data Mining Software	Analysing lake conditions and shoreline changes
18	Kara-Bogaz-Gol	Lagoon	Distributed Data Platform	Monitoring lagoon water quality and features
19	Soda Lakes of Kenya	Saline Lakes	Big Data Processing Tools	Analysing saline lake chemistry and sediments
20	Lake Assal	Crater Lake	Cloud-based Data Platform	Mapping crater lake chemistry and sediments

6.2.3 Hydrological Monitoring

Hydrological processes in wetland areas can be better understood with the help of remote sensing data. Researchers can learn more about wetland hydrology, water balance, and the effects of climate change or human activities on wetland water resources by evaluating surface water extent, water level fluctuations, and changes in wetland hydrological patterns.

6.2.4 Wetland Vegetation Monitoring

Wetland vegetation features including health, biomass, species composition, and dynamics can be monitored via remote sensing data like multispectral and hyperspectral photography. Invasive species can be pinpointed and the wetland's reaction to environmental changes can be monitored with the use of this data.

6.2.5 Wetland Ecosystem Functioning

The processes and functions of wetland ecosystems can be better understood with the help of remote sensing data. Wetland water storage capacities can be evaluated by using radar data to track soil moisture levels and thermal images to determine evapotranspiration rates.

6.2.6 Wetland Habitat Assessment

Wetland habitats can be evaluated, and ecological studies conducted, with the help of remote sensing data. It makes it possible to classify wetland landscapes according to their various habitat types, spatial patterns, and interconnectedness. Assessments of biodiversity, modelling of habitat suitability, and conservation planning can all benefit greatly from this data.

6.2.7 Wetland Restoration Monitoring

Wetland restoration progress can be tracked and evaluated with the help of remote sensing data. Changes in plant cover, hydrological conditions, and general wetland health can be evaluated by comparing images taken before and after restoration programmes were implemented.

6.2.8 Wetland Inventory and Database Creation

By providing reliable and spatially detailed information on wetland areas, distribution, and characteristics, remote sensing data can aid in the development of wetland inventories and databases. Wetland conservation strategies, policies, and decisions can all benefit greatly from the information provided by such inventories.

It is necessary to have access to suitable images, data processing tools, and analytical methods in order to make the most of remote sensing and Earth observation big data in wetland investigations. Wetland systems and their interactions with their environs can be better understood when remote sensing data is combined with other data sources, such as field measurements and socioeconomic data.

6.2.9 Climate and Weather Data

It is possible to learn about the climatic drivers and effects on wetland systems by analysing large data sets of climate and weather information such as historical records, real-time measurements, and the results of climate models. This information is useful for determining how susceptible wetlands are to the effects of climate change, examining hydrological patterns, and making predictions about the future.

The study of wetlands and their responses to climatic variability and change requires climate and meteorological data in the form of big data. Here are some key applications of climate and weather big data in wetland studies:

6.2.10 Climate Change Analysis

Changes in temperature, precipitation, and other climate variables can be evaluated over long time periods using climate data, illuminating the effects of climate change on wetland ecosystems. Wetland hydrology, vegetation, and general ecosystem dynamics may be impacted by future climatic scenarios, which can be understood by evaluating historical climate data and climate model forecasts.

6.2.11 Hydrological Modelling

Hydrological modelling of wetlands relies heavily on climate and meteorological data. Water balance components, surface and groundwater interactions, and wetland water levels and flows can all be estimated with the use of these data, which include precipitation, evapotranspiration, and prospective evapotranspiration. The accuracy of hydrological models can be increased by the use of big data analysis by incorporating large climate information.

6.2.12 Extreme Event Analysis

Extreme occurrences like floods, droughts, and storms can have profound effects on wetlands, and understanding these phenomena requires access to large amounts of data on climate and weather patterns. Wetland vulnerability and resilience can be evaluated with the help of climate model projections and historical weather records by analysing the frequency, intensity, and spatial distribution of extreme events.

6.2.13 Phenology and Vegetation Dynamics

The dynamics and phenology of marsh vegetation are difficult to analyse without access to climate data, such as temperature and precipitation. Researchers can learn about the impacts of climate change on wetland ecosystems by evaluating climatically driven phenological patterns like the timing of vegetation greening and senescence.

Climate-vegetation Interactions: Wetland vegetation relationships can be studied by combining climate and weather data with vegetation indices acquired from remote sensing. Climate-vegetation interactions, such as the impact of temperature and precipitation on wetland production, species composition, and distribution patterns, can be investigated through the use of big data analysis.

Climate Adaptation and Management: Wetlands management and climate adaption techniques can benefit greatly from the knowledge provided by climate and weather big data. Adaptive management strategies can be developed to lessen the effects of climate change on wetland ecosystems, and researchers can uncover potential implications by examining historical climate data and climate projections.

Climate Services and Decision Support: Big data from the fields of meteorology and climate science can aid in the management and preservation of wetlands. Future water allocation, land use planning, ecosystem restoration, and conservation activities can all benefit from accurate climate predictions made possible by climate services.

Climate Data Integration: Understanding the intricate interconnections within wetland systems is facilitated by combining climate data with other data sources such as hydrological data, ecological surveys, and socioeconomic information. Researchers can more accurately evaluate the combined effects of climate, hydrology, and human activity on wetlands if they combine data from a variety of sources.

Having access to trustworthy climate datasets, climate models, and data analysis tools is crucial for making good use of climate and weather big data in wetland studies. It is essential for multidisciplinary research and comprehension of the intricate dynamics between climate and wetland ecosystems that climatologists, hydrologists, ecologists, and wetland scientists work together (Holler et al. 2014).

6.2.14 Hydrological Data

Wetland hydrology may be better understood with the help of big data pertaining to hydrological processes such as river discharge, precipitation, water level, groundwater dynamics, and water quality measurements. Wetland hydrological regimes may be studied better by using these data to simulate water flow, analyse water balance, and evaluate the effects of human activities and climate change.

Understanding and analysing the hydrological processes and interactions of wetlands is greatly aided by hydrological data in the form of big data. Here are some key applications of hydrological big data in wetland studies:

Water Balance Analysis: The water balance of wetland systems can be analysed with the help of hydrological data including precipitation, evapotranspiration, groundwater levels, and streamflow. Researchers can better understand how wetlands function hydrologically if they can quantify the water that flows into and out of them.

Wetland Water Level Monitoring: Big data collected from stream gauges, groundwater wells, or remote sensing can shed light on the hydrodynamics of wetlands by showing how the water table rises and falls over time. Wetland flooding patterns, climate shifts, and seasonal shifts can all be better understood with the use of long-term records of water levels.

Groundwater-Surface Water Interactions: Groundwater and surface water interactions in wetlands can be evaluated with the help of hydrological data. The hydrological connectivity between wetlands and neighbouring aquifers, as well as analysing the effects of groundwater extraction or recharge on wetland water levels and ecosystem, depend on our ability to understand these interactions.

Flood and Drought Analysis: The incidence and effects of floods and droughts in wetlands can only be studied with the use of hydrological big data. Flood frequency, length, and intensity can be analysed with historical streamflow data and rainfall records, whereas prolonged periods of low flow or water shortages can be assessed with hydrological drought indices. These calculations aid in determining how susceptible or resistant a wetland is to floods and other severe water occurrences.

Water Quality Monitoring: The dynamics of wetland water quality can be better understood with the help of hydrological data and water quality measurements. To better understand water quality fluctuations and assess their effects on wetland ecosystems, it is important to monitor indicators such as nutrient concentrations, sediment loads, and pollutant levels in addition to hydrological data.

Hydrological Modelling: The creation and calibration of hydrological models, which can replicate water flow and dynamics in wetlands, rely heavily on hydrological big data. To simulate water flow, predict wetland water levels, and assess the effects of hydrological changes on wetland ecosystems, hydrological models use data inputs such as precipitation, evapotranspiration, soil properties, and land cover.

Climate Change Impact Assessment: The possible effects of climate change on wetland hydrology can be evaluated with the help of hydrological data. Researchers can assess how changes in precipitation patterns, temperature, and evapotranspiration might alter wetland water availability, flow regimes, and overall hydrological conditions by combining historical hydrological records with climate projections.

Integrated Water Resources Management: The management of water resources in and around wetlands relies heavily on hydrological big data. Sustainable water management strategies, efficient allocation of water resources, and the balancing of competing demands can all be achieved by taking into account the ecological requirements of wetland ecosystems through the integration of hydrological data with socio-economic information, land use data, and water demand projections.

The use of high-quality data, integration with other relevant data sources, and sophisticated analytical tools and models are all essential for making the most of hydrological big data in wetland research. The interpretation and application of hydrological data to inform wetland management, conservation, and decision-making requires collaboration among hydrologists, ecologists, water resource managers, and stakeholders.

6.2.15 Biodiversity Monitoring

Wetland biodiversity can be evaluated with the help of big data, including records of species occurrences and distributions as well as results from ecological surveys. Researchers can pinpoint biodiversity hotspots, monitor changes in species composition, and deduce what factors contribute to the decline or preservation of wetland biodiversity by studying large-scale species data.

The use of big data, which compiles vast amounts of information on species occurrences, distribution records, and ecological assessments, might considerably improve biodiversity monitoring in wetland studies. Here are some key applications of big data in biodiversity monitoring for wetland studies:

Species Occurrence and Distribution: The collection of species occurrence records in wetlands is aided by large data sources including citizen science projects, online databases, and biodiversity monitoring networks. Supporting the assessment of wetland biodiversity and conservation priority-setting, these records enable researchers discover species occurrence, distribution patterns, and hotspots of biodiversity.

Species Richness and Composition: The abundance and diversity of wetland species can be analysed with the help of big data. Researchers can analyse the diversity and abundance of different taxonomic groups, as well as the structure of communities and changes in species assemblages through time, by compiling records of species occurrences.

Rare and Endangered Species Monitoring: Rare and endangered wetland species are easier to spot and keep tabs on thanks to big data. Researchers can choose priority regions for protection, examine population trends, and quantify the success of conservation measures by examining species occurrence records and merging them with conservation status information.

Invasive Species Detection: Invasive species in wetlands can be detected and tracked with the help of big data. Wetland ecosystems are particularly vulnerable to invasion, thus scientists have combined species occurrence records with ecological niche modelling tools to anticipate the potential spread of invasive species and develop ways for early detection and management of these species.

Phenological Studies: By including observations of life-cycle events, including blossoming, migration, and reproduction, big data can aid phenological studies in wetlands. Researchers can learn more about the timing of these events, the extent to which they have been influenced by climate change and the range of possible consequences for wetland ecosystems and the species that live within them by studying phenological data at various spatial scales.

Species Interactions and Food Webs: The study of species interactions and food webs in wetlands is made possible by the use of big data analysis. Wetland ecosystems are notoriously complex, but by studying species co-occurrence and trophic interactions, scientists may better understand species interdependence and evaluate the robustness of wetland food webs.

6.2.16 Monitoring Ecosystem Health

The monitoring of wetland ecosystem health and function is aided by big data. The linkages between biodiversity and ecosystem health may be evaluated, significant causes of biodiversity loss can be identified, and wetland management methods can be informed by combining data on biodiversity with information on water quality, habitat features, and land use.

Data Integration and Collaboration: Monitoring biodiversity with big data requires researchers, organisations, and citizen scientists to work together and integrate their data sets. Big data may be used more effectively, data quality and coverage improved, and more complete wetland biodiversity monitoring supported if researchers pool data from different sources, standardise data formats, and build data-sharing platforms.

Strong data management, quality control, and cutting-edge analytical tools for data processing and interpretation are essential for using big data in wetland biodiversity monitoring. Data collection, validation, and the establishment of standardised techniques necessitate the involvement of researchers, conservation organisations, and citizen scientists. Ultimately, the use of big data in biodiversity monitoring aids in our comprehension of wetland ecosystems directs conservation efforts, and aids in the long-term sustainability of wetland biodiversity management.

Citizen Science and Crowdsourcing: Volunteers can help generate big data by reporting on wetland observations, wildlife sightings, and water quality measures using crowdsourcing platforms and citizen science programmes. Wetland research can benefit from the increased spatial and temporal coverage provided by these data sources, and more people can be involved in conservation efforts as a result.

Wetland research can benefit greatly from the large amounts of information gathered through citizen science and crowdsourcing projects, which encourage the general public to participate in data collection, analysis, and submission. Here are some ways in which citizen science and crowdsourcing can generate big data for wetland studies:

Species Observations: Wetland species, such as plants, birds, insects, and other organisms, can be tracked through citizen science platforms and apps. These observations are helpful for wetland biodiversity monitoring and study because they add to large-scale records of species occurrences, ranges, and phenology.

6.2.17 Education and Outreach

Opportunities for public participation, environmental education, and wetland conservation awareness are made possible through citizen science and crowdsourcing projects. These programmes instil a sense of responsibility and give individuals a voice in wetland study and preservation by having them participate in data gathering and research activities.

There needs to be strong data management systems in place, data quality control mechanisms put in place, and clear protocols and training provided for volunteers if wetland studies are going to make use of citizen science and crowdsourced data. In order to ensure data uniformity, correct potential biases, and maximise the scientific value of the supplied data, collaboration between scientists, citizen science project coordinators, and participants is vital.

Social and Economic Data: Human activity, socioeconomic indicators, land use/land cover changes, and policy data all make up part of the "big data" that can be used in wetland studies. This information is useful for assessing the efficiency of conservation efforts and learning more about the social and economic factors that influence wetland change.

Big data containing information about society and economy can shed light on the societal aspects of wetland research. Effective management, conservation, and decision-making hinge on an appreciation of the social and economic factors related to wetlands. Here are some ways in which social and economic big data can contribute to wetland studies:

Land Use and Land Cover: Economic and social data can reveal patterns in land use and land cover close to wetlands. Satellite imaging, remote sensing data, and GIS are all examples of big data sources that can be mined for information about the potential impacts of urbanisation, agriculture, and infrastructure development on wetland ecosystems.

Demographic Analysis: Social data, including population distribution, density, and demographic characteristics, help researchers understand the human population dynamics in the vicinity of wetlands. Demographic data can provide insights into population growth, urbanisation trends, and potential impacts on wetland ecosystems due to changes in land use or resource utilisation. Land use and land cover trends in the vicinity of wetlands can be determined with the aid of socio-economic data. Information about urbanisation, agricultural activities, infrastructure, and other land use changes that may affect wetland ecosystems can be gleaned from big data sources like satellite photography, remote sensing data, and GIS.

Socioeconomic Surveys: Wetland-related livelihood activities, economic dependencies, recreational use, and cultural values can all be gathered through surveys and questionnaires. The large amounts of information gathered from these surveys can be used to determine the value of wetlands to individual communities, pinpoint potential sources of social and economic instability, and investigate the connection between human actions and wetland preservation.

Economic Valuation: Wetland ecosystem services can be estimated using economic data. Researchers can calculate the monetary value of wetland services including water distribution, flood control, tourism, and recreation by evaluating large datasets of market prices, willingness-to-pay surveys, and cost-benefit assessments.

Stakeholder Mapping and Engagement: Stakeholders in wetlands can be identified and mapped with the help of social data. Local communities, NGOs, government organisations, and industries can all be identified through big data analysis, along with their respective interests, values, and possible conflicts. Engaging stakeholders, working together, and making decisions through cooperation and participation will all benefit from this data.

Environmental Justice and Equity: Wetlands' social and economic data can be utilised to examine questions of environmental justice and equity. By analysing large amounts of data, experts can determine whether or whether marginalised communities and vulnerable populations are disproportionately affected by or denied access to wetland resources and services.

Policy and Governance Analysis: Wetland policy and governance analyses can be aided by socioeconomic data. Analysis of large datasets can be used to determine whether or not current wetland management practices are producing the desired results, to evaluate regulatory frameworks, and to spot areas where social and economic components of governance should be strengthened.

Risk and Vulnerability Assessments: Assessments of risk and vulnerability in wetland environments can benefit from socioeconomic data. Changes in wetland ecosystems, such as flooding, habitat loss, or water pollution, can have a significant social and economic impact, and big data analysis can help pinpoint those at risk.

Interdisciplinary partnerships are necessary for making full use of social and economic big data in wetland research. Wetland management and decision-making must take into account both ecological and human dimensions, so researchers, social scientists, economists, and policymakers must collaborate to gather and analyse relevant data, integrate social and economic dimensions with ecological data, and develop comprehensive frameworks (Mayer-Schönberger and Cukier 2013).

6.2.18 Data Analytics and Machine Learning

The processing, analysis, and interpretation of huge and complicated datasets are made possible by big data analytics approaches, such as machine learning algorithms. In order to better understand and anticipate wetland dynamics and inform management decisions, these methods can be applied to the data to detect patterns, trends, and correlations.

Big data can be used in wetland research with the use of data analytics and machine learning methods to glean useful insights, discover patterns, and generate forecasts. Some examples of using data analytics and machine learning on large datasets for wetland research are shown below.:

Data Exploration and Visualisation: Large and complicated datasets pertaining to wetlands can be explored and visualised with the use of data analytics. Scatter plots, heatmaps, and GIS mapping are all examples of visualisation methods that can be used to examine and make sense of wetland data by highlighting geographical and temporal patterns.

Pattern Recognition and Classification: Wetland data can be analysed and categorised with the help of machine learning algorithms. Wetland habitats may be grouped together, vegetation groups can be identified, and water quality can be categorised depending on a number of different factors using algorithms like clustering and classification models. This can aid in the process of identifying, categorising, and monitoring wetlands.

Predictive Modelling: Predictive modelling in wetland research can be aided by machine learning. Models for predicting future wetland conditions, such as water levels, vegetation changes, or habitat suitability, can be constructed by training algorithms using historical wetland data, such as hydrological variables, land cover, or climatic data.

Species Distribution Modelling: Wetland species occurrence data can be used with machine learning approaches to create species distribution models. Models can predict species distributions, identify suitable habitats, and analyse the possible implications of environmental changes on species' ranges by combining species occurrence records with environmental variables like climate, land cover, or hydrological characteristics.

Anomaly Detection: Data collected from wetlands can be analysed for anomalies. In order to detect anomalies in wetland ecosystems, such as sudden shifts in water quality, atypical plant growth, or unusual hydrological conditions, machine learning algorithms can examine temporal or spatial data patterns.

Data Fusion and Integration: Wetland research can benefit from the use of data analytics to combine and combine data from many sources. Machine learning algorithms can assist in the creation of comprehensive and integrated datasets that provide a more in-depth understanding of wetland dynamics and interactions by merging disparate data sources like as remote sensing data, hydrological data, and socioeconomic data.

Uncertainty Analysis: Uncertainties in wetland data can be quantified and analysed with the help of data analytics techniques. Wetland modelling and analysis have its limitations, but machine learning methods can help evaluate the reliability and robustness of forecasts, pinpoint the causes of uncertainty, and shed light on those topics.

Data-driven Decision Support: Wetland management and conservation can benefit from data-driven decision-making enabled by data analytics and machine learning. Policymakers, land managers, and stakeholders can benefit greatly from the insights gleaned through predictive models, risk assessments, and scenario studies when it comes to wetland ecosystem management and conservation efforts.

Data quality, appropriate preprocessing, and feature engineering approaches are crucial for the successful application of data analytics and machine learning to large data in wetland studies. Wetland studies have unique goals and contexts, thus it's important that data analytics results and insights are developed in close partnership with domain experts, data scientists, and wetland researchers.

Data Integration and Collaboration: One of the most important aspects of using big data in wetland research is integrating datasets from a variety of different sources. In order to fully comprehend wetland systems, it is necessary to integrate information from numerous fields, including hydrology, ecology, remote sensing, and socioeconomics. To fully realise the promise of big data in wetland research, cooperation between researchers, organisations, and data providers is essential for data sharing, standardisation, and harmonisation.

Leveraging big data for wetland research relies heavily on data integration and collaboration. Here's how data integration and collaboration can enhance the use of big data in wetland research:

Data Sharing and Standardization: For data integration to be successful, information from many sources must be shared and standardised. Establishing data-sharing agreements, defining common data formats, and building data standards that enable smooth integration of varied datasets all need cooperation between academics, organisations, and stakeholders.

Multi-source Data Integration: Data from several sources, including satellite imagery, field surveys, citizen science projects, and official government records, must often be combined in order to fully understand wetlands. By working together, various data providers and stakeholders can more easily combine their data, leading to a more complete picture of wetland dynamics.

Cross-disciplinary Collaboration: Ecology, hydrology, sociology, and economics are only few of the fields that contribute to our understanding of wetlands. Cross-disciplinary methods for analysing large data are fostered through collaborative activities that bring together scholars from several fields. Big data analytics and the study of wetland ecosystems benefit from the incorporation of interdisciplinary knowledge and expertise.

Data Harmonisation and Interoperability: The integration of big data calls for the harmonisation and interoperability of disparate data sets. Wetland research can benefit greatly from the integration and analysis of big data if data providers and researchers work together to create data harmonisation methods, metadata standards, and interoperable systems.

Quality Control and Data Validation: Integrating big data requires a team effort to control and validate data quality. When various parties are involved, data gaps can be found, data quality can be evaluated, and validation methods can be put into place to guarantee the integrity of linked datasets. The ability to spot and fix data problems is another benefit of working together.

Advanced Analytical Approaches: The use of sophisticated analytic methods on massive data sets in wetland research is greatly aided by teamwork. Collaborative endeavours can leverage advanced data analytics methods like machine learning, data mining, and spatial modelling by combining their resources, experience, and computer capacity.

Data Privacy and Security: Addressing data privacy and security concerns is essential for effective collaboration in big data integration. Data suppliers, researchers, and institutions can work together to create privacy-preserving protocols and frameworks that guarantee compliance with privacy laws and facilitate safe data exchange.

Knowledge Sharing and Capacity Building: Researchers, practitioners, and stakeholders in the study of wetlands benefit from working together since it facilitates the exchange of information and the development of expertise. Training programmes, workshops, and conferences are all examples of collaborative efforts that can help researchers share knowledge and experience about using big data for wetland studies.

Integrating and sharing data requires parties to communicate, create trust, and coordinate their activities. Collaboration in the use of big data for wetland research can be improved by establishing data governance frameworks, data management procedures, and data access policies. Ultimately, cooperation improves everyone's knowledge of wetland ecosystems and facilitates decision-making in wetland conservation and management that is grounded in evidence.

Careful data management, data quality assessment, and the right analytical methods are essential for using big data in wetland research. It's also important to think about how to balance privacy with data acquisition. Big data can help scientists and policymakers learn more about wetland ecosystems, make better decisions based on facts, and improve wetland protection and management.

6.3 Internet of Things (IoT)

The Internet of Things (IoT) is a game-changing idea that has been gaining traction in recent years, changing the way we interface with both the online and offline worlds.

The term "Internet of Things" (IoT) is used to describe the global system of physical things, gadgets, and sensors that are linked together and exchange data over the Internet. Items as diverse as smart thermostats and wearable fitness trackers can be included here, as might heavy machinery and vehicles. IoT is unique because it allows all of these things to exchange data and information with one another.

The ability to collect and analyse data from disparate sources is a driving force behind the expansion of the IoT since it allows for better-informed decisions and greater productivity. Wearable Internet of Things devices are being used to keep tabs on patients' health and spot potential problems before they become serious. IoT applications control traffic, lessen energy use, and increase security in "smart cities". Many sectors will be affected by the Internet of Things. Internet of Things sensors improve crop productivity by allowing farmers to keep tabs on soil and plant health. Manufacturing operations benefit from IoT's real-time monitoring and predictive maintenance thanks to IoT. Internet of Things enables asset tracking, which improves supply chain management in logistics.

In addition, IoT is vital to the development of the modern customer experience. Lights, thermostats, and security systems in smart homes may all be managed from afar using Internet of Things gadgets. The Internet of Things is being used by stores to improve customer service and stock control. In the IoT, security and privacy are of utmost importance. Strong security mechanisms and encryption methods are needed to protect the large amounts of data created and shared by IoT devices from hackers.

Interoperability and standardisation are other challenges. The IoT ecosystem is diverse, with devices and platforms from various manufacturers. Ensuring seamless communication and compatibility among these devices is essential for the technology's growth.

Internet of Things is a groundbreaking technology that is reshaping our world by connecting the physical and digital realms. Its potential for data-driven insights and automation is driving innovation in various sectors. As IoT continues to evolve, it is imperative to address security, privacy, and interoperability concerns to harness its full potential for the benefit of society. Some examples of how the Internet of Things could be used in wetland research are also diagrammatically represented in Fig. 6.3 and some examples in Table 6.5 for wetlands and in Table 6.6 for saline wetlands.

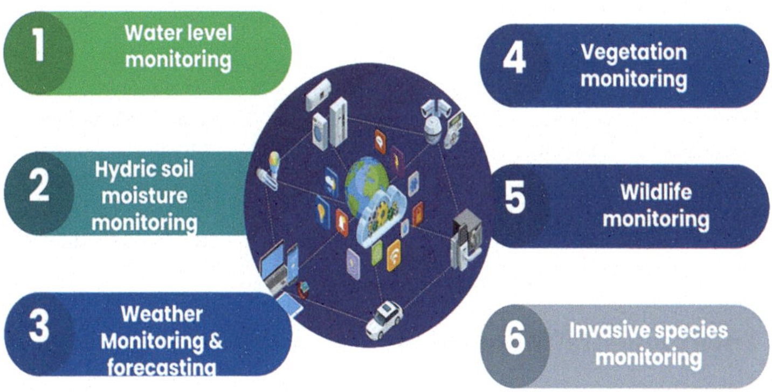

Fig. 6.3 Use of IoT for wetland studies

Table 6.5 Use of IoT for wetlands

S. no.	Name of wetland	Type of wetland	IoT technology used	Purpose of use
1	Everglades	Marshland	Weather Stations	Monitoring weather conditions and water levels
2	Okavango Delta	Inland Delta	Water Quality Sensors	Measuring water quality and pollution levels
3	Pantanal	Seasonal Floodplain	GPS Tracking Devices	Tracking wildlife movements and migration
4	Sundarbans	Mangrove Forest	Camera Traps	Monitoring tiger populations and behaviour
5	Danube Delta	River Delta	Water Level Sensors	Measuring water levels and flow rates
6	Kakadu National Park	Wetland	Soil Moisture Sensors	Monitoring soil moisture and vegetation
7	Hula Valley	Marshland	Bird Tagging Systems	Studying bird migration and habitats
8	Tonle Sap Lake	Flooded Forest	Water Quality Monitors	Assessing water quality and nutrients
9	Lake Chad	Endorheic Lake	Fish Tracking Devices	Tracking fish movements and populations
10	Okefenokee Swamp	Swamp	Remote Cameras	Studying swamp wildlife and behaviour
11	Liaohe Delta	River Delta	Water Flow Sensors	Measuring river flow and sediment transport
12	Narew National Park	Wetland	Environmental Sensors	Monitoring wetland conditions and water levels
13	Tisza River Basin	River Basin	Water Quality Monitors	Assessing river water quality and pollutants
14	Morava River Floodplain	Floodplain	Water Level Sensors	Measuring floodplain water levels and flow
15	Esteros del Ibera	Wetland	Wildlife Tracking Devices	Tracking wildlife movements and behaviour
16	Poyang Lake	Seasonal Lake	GPS Buoys	Monitoring lake water levels and currents
17	Rann of Kutch	Seasonal Salt Marsh	Soil Salinity Sensors	Measuring salt marsh salinity and conditions
18	Itaipu Dam Reservoir	Reservoir	Water Quality Monitors	Assessing reservoir water quality and pollutants
19	Humber Estuary	Estuary	Environmental Sensors	Monitoring estuary conditions and water quality
20	Lake Tahoe	Alpine Lake	Weather Stations	Studying lake weather conditions and climate

6.3.1 Environmental Monitoring

Wetland areas are ideal places to deploy IoT devices, as they can monitor several environmental factors in real time. Temperature, pH, dissolved oxygen, and nutrient levels are only a few of the water quality indicators that can be monitored by

Table 6.6 Use of IoT for saline wetlands

S. no.	Name of saline wetland	Type of saline wetland	IoT technology used	Purpose of use
1	Chott el Jerid	Salt Flat	Soil Salinity Sensors	Monitoring salt flat salinity and conditions
2	Sambhar Lake	Salt Lake	Water Quality Monitors	Measuring lake water quality and pollutants
3	Great Salt Lake	Terminal Lake	GPS Buoy Systems	Tracking lake water levels and currents
4	Lake Urmia	Endorheic Lake	Environmental Sensors	Monitoring lake area and environmental data
5	Maranjab Salt Desert	Salt Desert	Weather Stations	Monitoring desert weather conditions
6	Etosha Pan	Salt Pan	Wildlife Tracking Devices	Tracking wildlife movements and behaviour
7	Lake Natron	Alkaline Lake	Remote Cameras	Studying lake and shoreline wildlife
8	Salar de Uyuni	Salt Flat	Water Level Sensors	Measuring salt flat water levels and flow
9	Lake Eyre	Salt Pan	Bird Tracking Systems	Studying bird migration and habitats
10	Lop Nur	Salt Pan	Water Flow Sensors	Measuring pan water flow and sediment transport
11	Badwater Basin	Salt Pan	Soil Moisture Sensors	Monitoring pan soil moisture and conditions
12	Lake Abert	Endorheic Lake	GPS Tracking Devices	Tracking fish movements and populations
13	Lake Tuz	Salt Lake	Environmental Sensors	Assessing lake water quality and pollutants
14	Koyashskoye Salt Lake	Salt Lake	Weather Stations	Monitoring salt lake weather and climate
15	Sebkha Sejoumi	Salt Pan	Water Quality Monitors	Measuring salt pan water quality and pollutants
16	Lake Tengiz	Salt Lake	Weather Monitoring Stations	Analysing lake weather conditions and climate
17	Lake Qarun	Salt Lake	Fish Tracking Devices	Tracking fish movements and behaviours
18	Kara-Bogaz-Gol	Lagoon	Water Quality Monitors	Assessing lagoon water quality and pollutants
19	Soda Lakes of Kenya	Saline Lakes	Soil Salinity Sensors	Monitoring saline lake salinity and conditions
20	Lake Assal	Crater Lake	GPS Buoy Systems	Tracking lake water levels and currents

sensors. Rainfall, humidity, wind speed, and sun radiation are only some of the meteorological parameters that can be tracked by weather stations equipped with IoT sensors. Wetland dynamics can be better understood, and changes or disturbances can be addressed, with this round-the-clock monitoring.

Through the use of sensors and other devices connected to the IoT, wetland habitats can be monitored for a variety of environmental factors. Here are some examples of how IoT can be utilised for wetland environmental monitoring:

Water Level Monitoring: Changing water levels in wetlands can be tracked over time with the help of IoT water level sensors. Wetland hydrology, water balance, and the likelihood of floods can all be better understood with this information.

Water Quality Monitoring: In wetland water bodies, Internet of Things sensors with probes can assess factors like temperature, pH, dissolved oxygen, conductivity, and nutrient levels. In order to determine the state of wetland ecosystems, pinpoint the origins of pollution, and better manage water resources, constant monitoring of water quality is essential.

Soil Moisture Monitoring: IoT soil moisture sensors can be used to determine the soil moisture level in wetlands. Wetland hydrological processes, water management, and the effects of climate change can all be better understood with the help of soil moisture monitoring.

Weather Monitoring: Precipitation, wind speed and direction, air temperature, and humidity levels are only some of the data that may be collected by IoT weather stations set up in wetlands. This information is useful for analysing regional climate, discerning climatic trends, and gauging the impact of climate on wetland dynamics.

Vegetation Monitoring: Internet of Things devices, such as cameras and spectral sensors, can be used to keep an eye on the plant life in wetland areas. Photos and videos taken by IoT-enabled cameras can be used to assess shifts in wetland plant communities as well as their phenology and canopy cover. Spectral sensors can detect vegetation indices like the NDVI to learn more about plant health and productivity.

Wildlife Monitoring: Wildlife in wetlands can be monitored with IoT-based sensors like acoustic devices and motion-sensing cameras. Researchers can keep an ear out for amphibians, birds, and other vocalising animals in the marsh thanks to acoustic sensors. Images and videos taken by Internet-connected cameras can be used to evaluate species richness, behaviour, and habitat utilisation in wetlands.

Water Flow Monitoring: Stream and channel water flow rates in wetlands can be monitored with the use of IoT flow metres and velocity sensors. Wetland hydrodynamics, water availability, and the effects of flow changes on wetland habitats can all be better understood with the use of water flow monitoring.

Data Transmission and Management: IoT gadgets allow sensors to send data in real time to cloud-based servers or databases. Because of this, researchers and stakeholders are better able to access and comprehend the acquired data for informed decision-making through efficient data management, analysis, and visualisation.

Using IoT, researchers can keep tabs on the state of wetland ecosystems in real time. The information provided here is crucial for the conservation, management, and protection of the world's precious wetlands.

Hydrological Monitoring: IoT sensors like water level monitors, flow metres, and soil moisture monitors can provide real-time information on the hydrological state of wetlands. Water levels, water flow, groundwater levels, and soil moisture content can all be tracked with the help of such instruments when placed in wetlands. Water balance, hydrological processes, and the effects of climate change and land use can all be evaluated with the information gathered by IoT sensors in wetland areas.

The use of IoT sensors and devices for hydrological monitoring in wetlands entails gathering information on a wide range of wetland-specific hydrological characteristics. Here are some examples of how IoT can be utilised for wetland hydrological monitoring:

Water Level Monitoring: Changing water levels in wetlands can be continuously monitored with the use of IoT water level sensors. These sensors can report changes in water levels, tides, or precipitation/evaporation rates in near real time. Wetland hydrodynamics, water balance, and flood risk assessment can all benefit from this data.

Water Flow Monitoring: Streams, channels, and drainage systems found in wetland areas are ideal locations for installing IoT flow metres and velocity sensors. Wetland water movement may be evaluated, flow patterns identified, and the dynamics of water exchange with adjacent areas better understood through constant monitoring of water flow.

Groundwater Monitoring: Connected groundwater sensors can be placed in wetland wells or piezometers to monitor water table changes. In order to comprehend wetland hydrology, water availability, and the interplay between surface water and groundwater, monitoring groundwater dynamics is essential.

Precipitation Monitoring: Rainfall patterns can be studied with the help of IoT rain gauges or weather stations set up in wetland areas. Studying the effects of climate variability on wetland ecosystems requires knowledge of water supplies, the link between precipitation and wetland hydrology, and so on (McEwen and Cassimally 2013).

6.3.2 Wildlife Tracking and Monitoring

Wildlife in wetland environments can have IoT-based tracking devices attached to them, such as GPS tags and radio collars. Researchers can track animal whereabouts, behaviour, and habitat use with the help of these tools. Researchers can learn more about species distribution, migration patterns, and ecological connectivity in wetland ecosystems by studying the actions of wildlife and how they interact with wetland environments.

Using IoT-enabled devices and sensors, we can monitor wildlife in wetlands and keep tabs on their whereabouts, activities, and habitat preferences. Here are some examples of how IoT can be used for wildlife tracking and monitoring in wetlands:

GPS Tracking: Wildlife such as birds, animals, and reptiles can have IoT-based GPS trackers attached to them to monitor their whereabouts in wetland ecosystems. Research into animal movement patterns, critical habitats, and wetland connectivity can all be aided by the data collected and transmitted by these sensors at regular intervals.

Radio Telemetry: Radio collars or tags enabled by the IoT can be used to monitor wetland animals. These gadgets broadcast signals that can be picked up by IoT base stations or receiver networks, revealing the whereabouts of tagged animals in near-real time. Researchers can utilise this information to learn more about the ranges, migration routes, and habitat preferences of wetland species.

Acoustic Monitoring: In order to study the noises made by various wetland-dwelling animals, acoustic sensors based on the IoT can be set up in these areas. The ability of these sensors to recognise species-specific sounds or vocalisations sheds light on species richness, behavioural patterns, and the state of wetland ecosystems as a whole.

Camera Traps: Wildlife in wetlands can be monitored and photographed or filmed using Internet of Things-enabled camera traps. These cameras record day and nighttime wildlife behaviour when activated by motion or heat. Researchers are able to keep tabs on wetland species and their activity remotely thanks to the IoT's ability to transmit photographs in real time.

Environmental Sensors: Wetland wildlife can benefit greatly from the information provided by IoT devices outfitted with environmental sensors. Sensors monitoring water quality, for instance, can indicate whether or not circumstances are acceptable for aquatic species, while sensors measuring temperature, humidity, or light levels can help comprehend the microclimate within wetland habitats.

Data Transmission and Management: The IoT paves the way for real-time data transmission from animal tracking devices to cloud-based systems or databases. The acquired wildlife data can then be used for ecological studies, conservation planning, and adaptive management techniques thanks to improved data management, analysis, and visualisation.

6.3.3 Species Conservation and Management

The data collected through IoT-based wildlife tracking and monitoring can inform conservation efforts and management strategies for wetland species. It provides insights into habitat preferences, migration routes, and potential threats to wildlife populations. This information helps in designing targeted conservation initiatives, establishing protected areas, and implementing measures to mitigate human-wildlife conflicts. IoT devices make it possible for wildlife tracking data to be sent in real time to servers or databases. This improves the accessibility, processing, and visualisation of acquired wildlife data for ecological research, conservation planning, and adaptive management.

Researchers can learn more about species' movements, behaviours, and habitat use in wetlands thanks to IoT-enabled tracking and monitoring. Our knowledge of wetland ecosystems, conservation initiatives, and wildlife population management are all improved by the information provided here.

Remote Sensing and Imaging: Wetland research can benefit from the Internet of Things since it can improve remote sensing and imaging. High-resolution photographs and data of wetland areas can be captured by UAVs or satellites with sensors enabled by the IoT. Wetland vegetation dynamics, land cover changes, and boundary mapping are all areas where these tools may shine. Wetland ecosystems may now be monitored and analysed quickly and effectively because of the availability of real-time data transfer.

Integrating IoT devices and sensors with remote sensing technologies to gather and analyse data on wetland characteristics, dynamics, and changes is what is meant by remote sensing and imaging in wetlands utilising IoT. Here are some examples of how IoT can be used for wetland remote sensing and imaging:

Satellite Imagery: Satellite images of wetlands may be accessed and analysed using IoT devices. Researchers can automatically download and analyse satellite data to track wetland areas, vegetation cover, land use changes, and other metrics of interest by using IoT capabilities. This allows for widespread, systematic, and periodic monitoring of wetland ecosystems.

UAVs: UAVs can be made more effective for wetland remote sensing using the IoT. High-resolution data on wetland properties can be collected by equipping IoT-enabled UAVs with a variety of sensors, including multispectral or thermal cameras, LiDAR, or hyperspectral sensors. Researchers can keep an eye on wetland conditions in near-real-time thanks to UAVs and IoT.

Sensor Networks: Water quality, soil moisture, temperature, and vegetation indices are just some of the environmental characteristics that may be monitored with IoT-based sensor networks in wetland areas. Incorporating these sensors with remote sensing methods allows for the collection of continuous and spatially dispersed data on wetland conditions. A deeper comprehension of wetland dynamics is possible through the integration of data from remote sensing and IoT devices.

Image Analysis and Processing: IoT can be used to automatically analyse and analyse remote sensing photos. Researchers can evaluate and extract useful information from satellite or UAV pictures, such as wetland vegetation indices, land cover classifications, or change detection, by combining IoT capabilities with image processing techniques. This makes it possible to quickly and accurately retrieve wetland-related information from massive amounts of remote sensing data.

Data Fusion and Integration: Connecting remote sensing information with other data types, like in-person measurements and citizen science reports, is made easier by the Internet of Things. Researchers can improve the precision of wetland monitoring and assessments, authenticate remote sensing observations, and gain a more comprehensive understanding of wetland dynamics by merging data from numerous sources.

Data Visualisation and Decision Support: With the use of IoT, remote sensing data may be viewed and analysed in real time via online tools and platforms. The collected data can then be accessed and interpreted by stakeholders, researchers, and policymakers to guide their decisions about wetland conservation, management, and policy development.

Wetland monitoring and assessment at different spatial and temporal scales can be improved by combining IoT with remote sensing and imaging technology. Better conservation and management strategies for wetland ecosystems can be developed with the help of the information provided by this method.

Sensor Networks and Data Integration: Integration of sensor networks into research on wetlands is made easier by the Internet of Things. Researchers are able to obtain complete and synchronised data on characteristics including water quality, weather, and hydrological variables by placing a network of interconnected sensors throughout wetland areas. A more complete picture of wetland processes and interactions can be gleaned by combining data from various sensors (Michalski et al. 2013; Mukhopadhyay and Suryadevara 2014).

IoT sensor networks and data integration in wetland ecosystems entail the use of networked sensors and devices to monitor and record changes in a wide range of environmental factors. These sensors collect information on the hydrology, water quality, weather, and biodiversity of wetlands. Here's how sensor networks and data integration of IoT can benefit wetland studies:

6.3.4 Sensor Deployment

In order to collect data from widely dispersed locations in wetlands, sensor networks based on the Internet of Things can be deployed strategically. Water level sensors, soil moisture sensors, weather stations, water quality sensors, sound level metres, and wildlife tracking equipment are all examples of the types of sensors that can be used. These sensors work nonstop to compile information that paints a full picture of the state of wetlands.

Real-time Data Collection: Connected sensors in wetlands provide instantaneous data collection and analysis. Wetland conditions, hydrological changes, and environmental indicators can be continuously monitored by researchers and stakeholders. Quick decisions can be made with the help of real-time data, which is especially useful during emergencies such as floods, droughts, or pollution problems.

Data Integration: Information gathered by IoT sensor networks comes from a wide variety of places and measures. This involves combining information from numerous sensors, satellite images, and weather reports. By combining these data sets, scientists are able to examine intricate interrelationships and feedbacks within wetland ecosystems.

Data Management and Storage: Data may be stored and managed effectively on IoT platforms. Data is collected and stored in reliable, easily accessible locations like central databases or the cloud. This allows for long-term monitoring and analysis, data reusability, and improved communication among researchers.

Data Analysis and Visualisation: Integrating IoT data paves the way for cutting-edge methods of data analysis and visualisation. Data collected by the IoT can be subjected to machine learning algorithms, statistical analyses, and data visualisation tools in order to reveal trends, patterns, and outliers in wetland dynamics. This aids in understanding, forecasting, and making decisions based on evidence.

Early Warning Systems: IoT sensor networks can be utilised to create early warning systems for wetland-related dangers by collecting and analysing data in real time. To reduce dangers, alerts might be set off by sudden shifts in water level or water quality metrics, for instance, informing relevant parties in a timely manner.

Adaptive Management: Adaptive management strategies for wetlands can be bolstered by the constant flow of data provided by sensor networks enabled by the Internet of Things. By keeping tabs on a number of different indicators, stakeholders may gauge the results of management initiatives, measure the results of restoration efforts, and adapt their approach accordingly.

Improved monitoring capabilities, a deeper understanding of wetland dynamics, and more well-informed conservation and management decisions are all possible thanks to sensor networks and IoT data integration in wetland studies.

6.3.5 Early Warning Systems

Wetland ecosystems may benefit from Internet of Things-based early warning systems that monitor for and report on potential dangers and threats. IoT sensors can track things like rainfall and water levels, sending out warnings if certain conditions are met that could lead to flooding. Proactive steps can be performed with the help of such systems, reducing the risk of harm to the wetland environment and allowing for prompt responses to safeguard the ecology and the surrounding communities.

Using the IoT, Early Warning Systems (EWS) for wetland ecosystems detect and respond to possible dangers or hazards through the use of sensor networks and data processing techniques.

Sensor Deployment: Sensor networks powered by the Internet of Things are deployed strategically around the wetland to monitor environmental factors like water level, water quality, weather, soil moisture, and biodiversity indicators in real time. Wetland dynamics can be fully comprehended because of the constant monitoring by these sensors.

Data Monitoring and Analysis: The information gathered by sensors on IoT devices is instantly transmitted and evaluated. Machine learning algorithms and other forms of advanced data analytics can be used to examine the data for anomalies, patterns, and trends. The EWS can detect possible dangers or hazards by comparing current conditions to predetermined criteria.

Risk Identification and Alert Generation. The EWS can detect threats to the wetland ecology by analysing collected data. Examples include alterations in biodiversity patterns, which may indicate ecological disturbances, and sudden changes in water levels, which may indicate floods or pollution occurrences. When such incidents are identified, the EWS sends out warnings or notifications to the appropriate parties.

Real-time Communication and Stakeholder Engagement: With an EWS built on the Internet of Things, you can interact with your stakeholders instantly. Mobile devices, emails, and websites are all viable options for disseminating alerts and notifications. As a result, stakeholders can be made aware of any potential dangers and allowed to take preventative measures swiftly.

Rapid Response and Mitigation Measures: The EWS provides early warnings of possible threats in wetlands, allowing stakeholders to take immediate action. In the event of flooding, for instance, rescue workers can be quickly dispatched, and evacuation preparations can be quickly put into action. Wetland ecosystems are less likely to be negatively affected if the causes of water quality problems are addressed.

Decision Support and Adaptive Management: The EWS's data can help with both planning and adjusting management strategies for wetland areas. To better conserve and restore wetland ecosystems, decision-makers can evaluate the efficacy of previous management measures by looking back at historical data and patterns.

Enhanced Resilience and Conservation: The resilience and preservation of wetlands benefit from the use of an EWS based on the Internet of Things. The EWS aids in disaster relief by alerting authorities to impending danger and facilitating quick action in the event of an emergency. As a result, wetland habitats are better protected and maintained for the future.

IoT is being used in early warning systems for wetland ecosystems and human populations to increase their ability to recognise and respond to possible threats, therefore mitigating their harmful effects. An IoT-based EWS improves the resiliency and management of wetlands in the face of changing environmental circumstances by leveraging real-time data collecting, processing, and communication.

6.3.6 Data Analytics and Decision Support

Data analytics and machine learning methods can be used to process and evaluate IoT-generated data. This facilitates the discovery of trends and outliers in wetland data, which in turn informs management and conservation decisions. Wetland conditions can be predicted for the future using adaptive management tactics thanks to predictive models built from the Internet of Things data.

Wetland management and conservation can gain a lot from IoT-based data analytics and decision support systems. Here's how IoT-based data analytics and decision support systems can be used in the context of wetlands:

Data Collection: Water levels, water quality indicators, weather conditions, soil moisture, vegetation indices, and wildlife activity are just some of the environmental data that may be collected using IoT devices and sensor networks in wetlands. These sensors operate nonstop, generating massive amounts of data in real time.

Data Integration: Data from sensors, satellite images, weather stations, and other data stores can all be combined with the help of the Internet of Things. By combining and analysing the acquired data, a thorough picture of wetland conditions and dynamics may be painted. Correlations, patterns, and relationships between variables can be better understood with the use of combined datasets.

Real-time Data Analytics: The Internet of Things makes real-time data processing possible, which speeds up both discovery and decision-making. Trends, anomalies, and patterns can be identified by applying sophisticated data analytics approaches to the gathered information, such as machine learning algorithms and statistical models. Wetland processes can be better understood using real-time analytics, allowing for more proactive management and more prompt interventions.

Predictive Analytics: Predictive models can be created using IoT-based decision support systems by assessing both historical and real-time data. Wetland conditions such as water levels, water quality trends, and ecological shifts can all be predicted with the use of these models. Wetland management techniques, conservation efforts, and resource allocation can all benefit from the use of predictive analytics.

Visualisation and Dashboarding: IoT-based decision support systems can visualise and dashboard the studied data for ease of use. These visualisations simplify complex data for readers and improve its readability for decision-makers. Wetland conditions can be visualised in a variety of ways, such as on a map, a chart, a graph, or through an interactive tool.

Scenario Planning and Risk Assessment: Wetland management scenario planning and risk assessment can benefit from IoT data analytics. The possible effects of various interventions or changes in wetland conditions can be evaluated through scenario modelling. Areas at risk can be pinpointed, the possibility of various threats (such as flooding or pollution) evaluated, and preventative measures prioritised with the help of risk assessment models.

Adaptive Management: Adaptive management strategies for wetlands are made possible by IoT-based decision support systems. The success of management initiatives can be tracked and evaluated with the help of continuous data collection, analysis, and feedback loops. Wetland management can be made more effective and sustainable when policymakers make adjustments based on real-time data.

Stakeholder Engagement: Stakeholder participation in wetland management can be improved with the help of IoT-based decision support systems. These platforms let academics, politicians, local communities, and conservation organisations communicate and work together by making data readily available and easy to interpret. Data sharing promotes openness and group decision-making.

Using Internet of Things-based decision support systems can increase participation from key stakeholders in wetland management. These technologies allow researchers, politicians, community members, and conservation groups to communicate and work together by delivering easily digestible data. When information is easily accessible to all, it encourages open dialogue and consensus building (Nilsson 2009; Russell 2010).

6.3.7 Citizen Science and Public Engagement

The public can be involved in wetland research with IoT devices by participating in citizen science projects. People can help gather information by using IoT devices or smartphone apps to monitor and report on wetland species, water quality, and other environmental aspects. Participation like this not only improves the quality of data collected but also increases public interest in and action toward wetland preservation.

Wetland protection, restoration, and study all benefit greatly from public participation and citizen science. The IoT has the potential to greatly improve citizen scientific efforts and public participation in wetland-related pursuits. Here's how IoT can be used for citizen science and public engagement in wetland studies:

Data Collection: Citizen scientists can be equipped with IoT devices like sensors or mobile applications to monitor and report on a wide range of wetland metrics. Water quality metrics such as temperature, pH, or dissolved oxygen can be measured by volunteers using Internet of Things-enabled water quality sensors. They can also use GPS and camera apps on their smartphones to record their location data and take photos of wetland characteristics.

Data Sharing and Crowdsourcing: Citizen science data about wetlands can be shared and crowdsourced with the help of Internet of Things-based platforms and applications. Researchers, managers, and other interested parties can access and evaluate uploaded data through a centralised database or cloud-based platform. Crowdsourcing inspires community involvement and paves the way for the collection of massive data sets from a variety of wetland sites.

Real-time Monitoring: Through the use of Internet of Things (IoT) devices and sensors, citizen scientists may help with ongoing wetland monitoring. During floods, for instance, IoT-enabled water level sensors can report current levels in real time. Rapid response and sound decision-making are bolstered by real-time monitoring's improved timeliness and quality of data collecting.

Public Education and Awareness: Thanks to the real-time monitoring capabilities made possible by IoT devices and sensors, citizen scientists can take part in the ongoing monitoring of wetland conditions. IoT-enabled water level sensors, for instance, can report on flood levels as they happen. Increased timeliness and precision in data collecting made possible by real-time monitoring facilitates prompt action and sound judgement.

Participatory Research and Co-creation: Through the Internet of Things, citizen scientists can take an active role in the research process. Researchers can enhance spatial coverage, gain access to unique viewpoints, and learn from local expertise when they invite the public at every stage of a study, from conception to analysis. Through this collaborative method, citizen scientists gain a sense of agency and ownership of their work.

Gamification and Incentives: Gamification components like challenges, leaderboards, and awards can be incorporated into IoT-enabled citizen scientific platforms to increase user interest and involvement. By making it more fun, gamification techniques increase the likelihood that people will continue to participate in wetland monitoring and conservation efforts over time.

Collaboration and Networking: Collaboration and networking among participants can be facilitated through citizen science projects based on the Internet of Things. Through online discussion groups, social media communities, and specialised platforms, citizen scientists are able to network with one another, share their experiences, and learn from one another. Cooperation helps people feel more connected to one another and offers possibilities for growth and development.

Policy Advocacy and Decision-making: Data collected through Internet of Things-enabled citizen science projects is useful for advocating for and making decisions about public policy based on evidence. The collected information can be used to raise awareness about the value of wetlands, pinpoint potential risks or degradation, and guide policy discussions at the regional, state, and federal levels.

Wetland research can gain from enhanced data collecting, public awareness, and collaborative efforts if the public is encouraged to participate through the use of IoT technology for citizen science and public engagement. The long-term survival of wetland ecosystems is aided by public involvement in monitoring and conservation efforts, which also instils a feeling of environmental responsibility, strengthens communities, and benefits local residents.

Data privacy, security, and network connectivity are all crucial factors that must be taken into account when applying IoT to wetland research. In addition, researchers, technologists, and stakeholders need to work together to successfully adopt IoT solutions and use the resulting data to enhance wetland management and conservation efforts.

References

Barrat J (2023) Our final invention: artificial intelligence and the end of the human era. Hachette UK, London

Chaudhuri A (2018) Internet of things, for things, and by things. CRC Press, Boca Raton

De Brouwer PJ (2020) The big R-book: from data science to learning machines and big data. John Wiley & Sons, New York

Greengard S (2021) The internet of things. MIT Press, Cambridge

Hassan QF, Madani SA (eds) (2017) Internet of things: challenges, advances, and applications. Routledge, London

Holler J, Tsiatsis V, Mulligan C, Karnouskos S, Avesand S, Boyle D (2014) Internet of things. Academic Press, New York

Mayer-Schönberger V, Cukier K (2013) Big data: a revolution that will transform how we live, work, and think. Houghton Mifflin Harcourt, Boston

McEwen A, Cassimally H (2013) Designing the internet of things. John Wiley & Sons, New York

Michalski RS, Carbonell JG, Mitchell TM (eds) (2013) Machine learning: an artificial intelligence approach. Springer Science & Business Media, Berlin

Mukhopadhyay SC, Suryadevara NK (2014) Internet of things: challenges and opportunities. Springer International Publishing, Berlin, pp 1–17

Nilsson NJ (2009) The quest for artificial intelligence. Cambridge University Press, Cambridge

Russell SJ (2010) Artificial intelligence a modern approach. Pearson Education, Inc., London

Part III

Saline Wetlands—Case Study

Inland Saline Wetlands

7

Abstract

This section offers an in-depth exploration of the distinct attributes and ecological importance of "Inland Saline Wetlands". These habitats, marked by elevated salinity levels, hold significant roles in fostering biodiversity, regulating climate, and purifying water. Through the presentation of case studies, readers gain insights into the complexities surrounding inland saline wetlands, including the challenges they face and the ongoing conservation initiatives aimed at preserving these invaluable ecosystems.

Keywords

Sambhar lake · Geology · Salinity · Biodiversity · Livelihood

Unique and biologically significant, inland saline wetlands can be found in many different parts of the world and are defined by the presence of wetland plants and the presence of brackish or salty water. These ecosystems are crucial because of the biodiversity they sustain, the ecosystem services they offer, and the positive impact they have on the environment.

Saltwater intrusion, naturally occurring salt flats, and geological conditions are often linked to the presence of saline wetlands in both coastal and inland settings. The interplay between hydrological processes and the presence of saltwater shapes them, often yielding a rich mosaic of habitat types.

There are several environmental benefits from preserving these wetlands. First and foremost, they are vital to the survival of a diverse variety of plant and animal species because they provide a home for organisms that have evolved to survive in salty circumstances but would perish in freshwater settings. Many species of birds, insects, and fish rely on these wetlands as essential habitats for their survival. As a

L. K. Sharma, R. Naik, *Conservation of Saline Wetland Ecosystems*,
https://doi.org/10.1007/978-981-97-5069-6_7

result of evolution, the flora and fauna of these areas are distinct and frequently extremely rare, adding to the diversity of life on Earth.

Inland saline wetlands also have considerable ecological functions. As natural filters, they assist in cleaning water before it runs into larger bodies of water by eliminating nutrients and impurities. They can also play a role in reducing the effects of climate change by acting as carbon sinks by storing carbon in their sediments and biomass.

These wetlands are also commercially and culturally significant. Historically, they were employed in the salt industry, in folk medicine, and as grazing land. They help boost tourism because of the attractions they provide for leisure activities. However, urbanisation, agriculture, and alterations in water management methods pose a threat to the long-term viability of these ecosystems by causing the loss and degradation of their habitats.

Inland saline wetlands are extraordinary and important ecosystems that serve important ecological and cultural roles. In order to keep biodiversity high, water quality high, and these amazing landscapes sustainable, their preservation and management are crucial. It is essential to strike a balance between nature and human communities by conserving and maintaining these ecosystems. The comparison between freshwater wetlands and saline wetlands is given in Fig. 7.1.

Salt Pans: When water evaporates, it leaves behind a flat, shallow depression or basin, which over time becomes a salt pan. They tend to populate dry or semiarid areas where precipitation is scarce and water is rapidly lost to evaporation. Adapted to extreme salinity, halophytic plants can thrive in salt pans and create unique salt crust forms.

Saline Lakes: Saline lakes are inland water bodies that contain high levels of salt. The accumulation of salts over time is what causes them to form in isolated basins where water comes in but nowhere else can go. Size and salinity of saline lakes range widely, with some harbouring specialised salt-loving fauna.

	Coastal wetlands	Inland saline wetlands
Ocean connection	Connected	Not connected
Geographical location	Coastal areas	Semi-arid & arid regions
Water level	Constant	Fluctuation
Salt concentration	35 gm/l	3 to 300 gm/l
Examples	Salt marsh, Mangrove swamps	Playas, Salt pans

Fig. 7.1 Comparison between coastal and inland saline wetland

Saline Springs: Freshwater springs with a high salinity level due to dissolved salts in the groundwater are called saline springs. They can be found in a wide variety of geological environments and frequently form tiny wetland habitats near the spring outlets. The unique flora and fauna of saline springs are suited to the salty water.

Saline Playas: Saline playas are transient wetlands that become inundated during wet seasons like rain or snowmelt. High salt concentrations result from evaporation and the absence of outflow, therefore they can even be found inland. Unique salt-tolerant plants and migratory bird habitats may thrive in inland salty playas.

Alkaline Wetlands: High amounts of carbonate minerals or bicarbonate ions in the water or soils are usually responsible for the high pH found in alkaline wetlands. Dissolved salts can cause these wetlands to become more salty than usual. The presence of carbonate-rich geology or alkaline groundwater can both contribute to the formation of alkaline wetlands.

Salt Flats: When shallow bodies of water evaporate, the salt that remains covers vast regions, creating what is known as "salt flats". They typically appear in dry or semiarid areas that have salty groundwater or unreliable water supplies. Salt flats can be home to rare populations of salt-tolerant plants and serve as critical avian and mammalian habitats (International Conference on Salt Lakes 1999; Denny 2012; Devlin et al. 2021; Scott et al. 2014; Kalff 2002; Jorgensen et al. 2012).

References

Denny P (ed) (2012) The ecology and management of African wetland vegetation: a botanical account of African swamps and shallow waterbodies, vol 6. Springer Science & Business Media, Berlin

Devlin A, Pan J, Shah MM, Nuhu AA (eds) (2021) Inland waters: dynamics and ecology. IntechOpen, London

International Conference on Salt Lakes (7, 1999) (2001) Saline lakes: publications from seventh international conference on salt lakes: held in Death Valley National Park, California, USA, September 1999. Kluwer, Riverwoods

Jorgensen S, Tundisi JG, Tundisi TM (2012) Handbook of inland aquatic ecosystem management, vol 5. CRC Press, Boca Raton

Kalff J (2002) Limnology: inland water ecosystems. Prentice Hall, New Jersey

Scott DB, Frail-Gauthier J, Mudie PJ, Mudie PJ (2014) Coastal wetlands of the world: geology, ecology, distribution and applications. Cambridge University Press, Cambridge

Case Study: Sambhar Lake, India

8

Abstract

This section takes us to the arid landscapes of India, where we examine the "Sambhar Lake" case study. Sambhar Lake is India's largest inland saline wetland and holds immense cultural and ecological value. By studying this case, we gain insights into the complex interplay of factors affecting the lake, from changing land use to the conservation measures in place. It serves as a real-world example of the critical balance required to protect and sustain such fragile ecosystems.

Keywords

Eco-tourism · Geology · Hydrology · Livelihood · Migratory birds

The state of Rajasthan in northwest India is home to Sambhar Lake, the country's largest saline lake and one of the most significant inland wetlands in the area. It's a large natural feature and important ecological hotspot, covering over 190 square kilometres.

Because of the dry climate, Sambhar Lake is notorious for its excessive salinity. Five freshwater rivers feed into it, but it has no way to drain, so over time the lake has become increasingly mineralised and salty. Because of its salty composition, the lake is home to a wide variety of avian life.

The lake serves as a critical breeding and wintering area for a number of bird species, making it one of its most remarkable features. Migratory birds, especially flamingos, find the area's shallow waters and expansive mudflats to be a suitable habitat for foraging and reproduction. Tens of thousands of flamingos flock to the lake throughout the winter, making it a fantastic destination for birdwatchers and nature lovers.

L. K. Sharma, R. Naik, *Conservation of Saline Wetland Ecosystems*,
https://doi.org/10.1007/978-981-97-5069-6_8

Sambhar Lake is also very important to the local economy. The salt pans that dot the landscape around its coastlines attest to its long history as a salt-producing hub. The local populace relies heavily on salt mining as the region's principal economic activity.

Because of the lake's rare ecology and vital role in the regional economy, solving its many problems has proven difficult. Pollution, habitat deterioration, and fluctuating water levels have damaged the fragile ecosystem of the lake, harming the resident and migratory bird populations. To combat these problems and safeguard the lake's ecological and cultural significance, sustainable management and conservation initiatives are required.

Sambhar Lake is an impressive natural landmark. Because of its ecological and economic significance, it is a centre of high salinity research and salt production as well as a haven for a diverse population of birds. Responsible conservation measures are essential to maintain the survival of this rare inland saline wetland in the face of mounting environmental threats (Fig. 8.1).

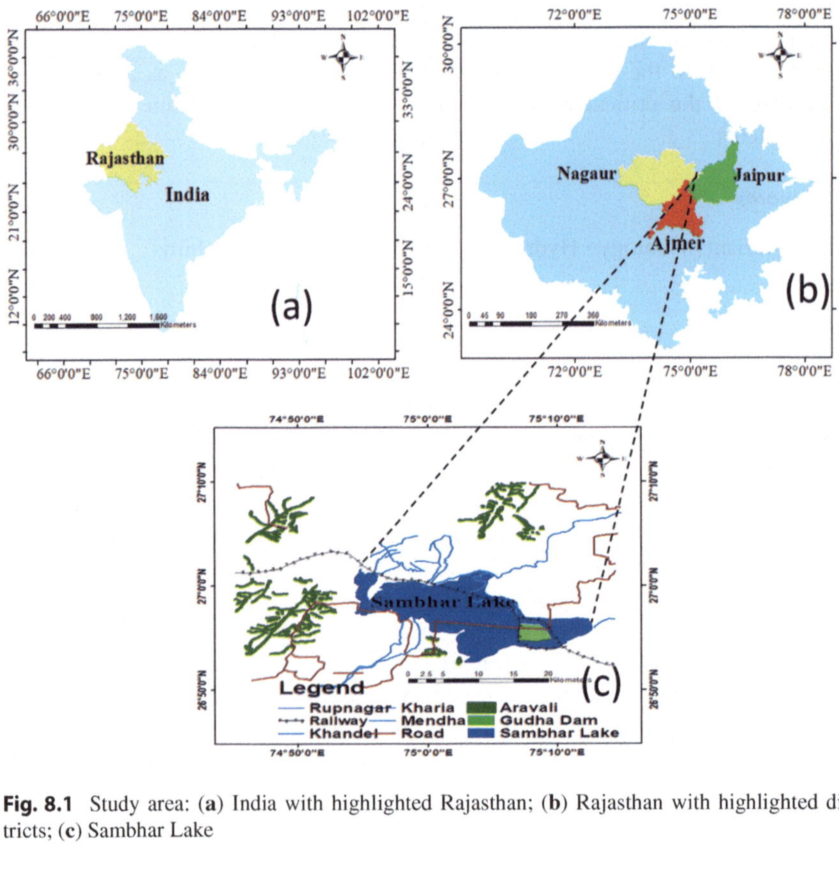

Fig. 8.1 Study area: (**a**) India with highlighted Rajasthan; (**b**) Rajasthan with highlighted districts; (**c**) Sambhar Lake

8.1 History and Civilisations

Sambhar Lake is the largest saltwater lake in India and among the largest in the world. The lake and its surroundings have been linked to many different cultures throughout history. Sambhar Lake has been home to many different cultures throughout time.

Ancient Period: There is proof of human presence near Sambhar Lake all the way back to the Harappan era (around 3000 BCE). Harappan items, including pots, knives, and other implements, have been discovered during excavations near the lake.

Maurya Empire: Ashoka, emperor of the Maurya Empire (321–185 BCE), brought Buddhism to the area and erected a monument honouring the Buddha near Sambhar Lake. Inscriptions in Brahmi writing can be seen in the Sambhar Lake Pillar.

Gurjara-Pratihara Dynasty: A powerful Rajput kingdom known as the Gurjara-Pratihara dynasty controlled the area in the eighth century CE. Around Sambhar Lake, they erected a number of forts and temples. The Pratihara emperors mined the lake for salt, a precious commodity back then.

Mughal Period: The Mughals recognised the strategic and commercial value of Sambhar Lake and its salt production facilities. Akbar and Jahangir, among other Mughal rulers, took a personal interest in the lake's administration because of its importance to the empire's salt supply.

British Raj: Sambhar Lake was once part of the princely state of Jaipur before the British took over India. The British expanded the salt works and included cutting-edge methods for creating salt (Bhat et al. 2015; British Library n.d.; Cherekar and Pathak 2015).

Post-Independence: Sambhar Lake and the surrounding salt fields fell under the jurisdiction of the Government of Rajasthan after India's independence in 1947. The lake is still a major hub for salt production, and many different salt businesses can be found there (Fig. 8.2).

Fig. 8.2 Sambhar Lake, 1875. (Image Source: British Library)

8.2 Geography

The Nagaur and Jaipur districts of Rajasthan, in the northwest of India, are home to Sambhar Lake. About 80 km to the southwest of the state capital, Jaipur. Some important details about Sambhar Lake's physical features are as follows:

Location and Size: Sambhar Lake can be found in the Great Indian Desert (Thar), popularly known as the Thar Desert. It lies between 26°54′N and 75°13′E in the coordinate system.

Formation: Sambhar Lake is an endorheic lake, meaning that it does not drain into the sea and is hence shallow. Over millions of years, geological processes in the area gave rise to its present appearance. The lake's distinctive salt flats are actually a combination of clay, sand, and salt that make up the lake bed.

Hydrology: The lake's water largely comes from rainfall and a few small streams that drain into it during the monsoon season. Concentrations of salt in the lake are between 8% and 10%, making it extremely salty. It plays a crucial role in India's salt industry.

Flora and Fauna: The brackish water of the lake and the adjacent marshland create a special environment that is home to many different kinds of plants and animals. Flamingos, pelicans, storks, and many species of ducks and waders use the lake as a wintering ground because of the abundance of aquatic vegetation.

Islands: Several islets dot Sambhar Lake's vast surface. Shakambari Devi Island, the largest of them, is named after the Hindu goddess of the same name. The island is sacred ground because it is home to a temple honouring the goddess.

Salt Production: For generations, salt has been harvested from Sambhar Lake. Several different salt firms manage the large salt pans that surround the lake. Evaporating lake water and collecting the resulting salt crystals is the conventional method of salt extraction.

8.3 Geomorphology

Sambhar Lake in Rajasthan is shaped by a number of geological processes and features, all of which contribute to the lake's geomorphology. Sambhar Lake's essential rock types are given in Table 8.1. Geomorphological features are as follows:

Salt Flats: Sambhar Lake's huge salt flats are its most striking geomorphic feature. The lake's striking white colour is due in large part to the extensive salt flats that cover much of the lake bed. Because of the high salt content of the lake water, salt flats are developed. Salt crystals form over time as water evaporates, leaving a coating behind.

Playas and Sabkhas: Sambhar Lake combines features of a playa and a sabkha. A playa is a shallow depression that is flat on the bottom and only becomes flooded when it rains. Sabkhas, on the other hand, are salt flats that develop along the

Table 8.1 Different rock types of Sambhar Lake

S. no.	Rock type	Chemical formula	Properties	Importance in Sambhar Salt Lake, India
1	Halite (rock salt)	$NaCl$	Sodium chloride; cubic crystals	Primary source of salt production in the lake
2	Gypsum	$CaSO_4{\cdot}2H_2O$	Hydrated calcium sulphate; white, translucent crystals	Occurs in evaporative deposits and is used in salt-making
3	Anhydrite	$CaSO_4$	Anhydrous calcium sulphate; harder than gypsum	Found in evaporite deposits and geological strata
4	Dolomite	$CaMg(CO_3)_2$	Carbonate rock; composed of calcium and magnesium	Part of the sedimentary formations in the lake
5	Shale	Varied composition	Sedimentary rock; fine-grained and layered	Part of the geological strata and sedimentary history
6	Sandstone	Varied composition	Sedimentary rock; composed of sand-sized grains	Geological history and formations around the lake
7	Limestone	$CaCO_3$	Carbonate rock; composed of calcium carbonate	Associated with the geological history of the area
8	Mudstone	Varied composition	Sedimentary rock; fine-grained and compacted	Common in the sedimentary formations of the lake
9	Siltstone	Varied composition	Sedimentary rock; composed of silt-sized grains	Part of the sedimentary history and geological strata
10	Schist	Varied composition	Metamorphic rock; foliated, often shiny	Occurs in the surrounding geological formations
11	Granite	Varied composition	Igneous rock; coarse-grained, often with visible minerals	Part of the surrounding geological landscape
12	Sand	Varied composition	Sediment; composed of fine to coarse grains	Contributes to the lake's sedimentary deposits
13	Lignite	Varied composition	Sedimentary rock; soft, brownish-black, carbon-rich	May be found in sedimentary formations near the lake
14	Basalt	Various compositions	Igneous rock; fine-grained, dark-coloured	Associated with the geological history of the region
15	Lepidolite (mica)	$K(Li,Al)_3(AlSi_3O_{10})(F,OH)^2$	Silicate mineral; mica group, often used in industry	Occurs in the surrounding geological formations
16	Slate	Varied composition	Metamorphic rock; fine-grained, easily split into layers	Part of the geological formations in the area

(continued)

Table 8.1 (continued)

S. no.	Rock type	Chemical formula	Properties	Importance in Sambhar Salt Lake, India
17	Sand dunes	Varied composition	Loose granular material; shifting dunes and sand hills	Affect the landscape and surrounding environment
18	Gneiss	Varied composition	Metamorphic rock; foliated and banded	Part of the geological formations in the region
19	Agate	SiO_2	Silica mineral; fine-grained, often with banded patterns	Occurs in the geological formations around the lake
20	Clays (various types)	Varied compositions	Sedimentary rock; fine-grained and malleable	Influence the sediment and soil characteristics

coast or further inland in arid climates. Playas and sabkhas form in various parts of the lake as a result of the lake's ebb and flow.

Alkali Soils: Soils in and surrounding Sambhar Lake tend to be very alkaline. Soluble salts like sodium, potassium, and magnesium can be found in abundance in these soils. The high salt content of the lake and the gradual evaporation of salty water over time have led to the development of alkaline soils in the area.

Evaporite Deposits: Sambhar Lake is a major evaporite deposit location. Minerals known as evaporites form when water is evaporated from solutions of a high enough concentration. Evaporation of the brackish lake water causes the dissolved salts to crystallise, leaving behind layers of evaporite. It is mined heavily for its halite (rock salt) because it is the most abundant evaporite mineral there.

Shoreline Features: There are sandbars, mudflats, and marshes all along the lake's shoreline. Water levels and sedimentation patterns affect the appearance of these features. The shoreline widens and the water level rises during the monsoon season, but the mudflats and marshes become more noticeable during the dry season.

Islands: Several islands add variety to the geomorphology of Sambhar Lake. Sedimentary materials like clay, sand, and salt are primarily what make up these islands. Shakambari Devi Island, the largest island in Sambhar Lake, is revered for its religious and cultural significance (Gupta et al. 2015; Jain et al. 2005; Jakher et al. 1990).

8.4 Climate

Sambhar Lake, in Rajasthan, faces climate conditions typical of the desert because of its proximity to the dry Thar Desert. Sambhar Lake's primary weather features are as follows:

Hot and Dry: The weather around Sambhar Lake is typically very warm and dry. The months of April through June are the height of summer, and temperatures regularly exceed 40 °C during this time. Sun and heat are at their peak around the lake and its surroundings during this time of year.

Monsoon Season: The region has a short monsoon season, lasting only from July to September. The monsoon provides welcome reprieve from the sweltering heat and helps to replenish the water in Sambhar Lake. However, annual precipitation is only about 200–300 mm (8–12 in.).

Winter Season: From November to February, Sambhar Lake has a warm and beautiful winter. The average temperature falls between 10 and 25 °C during this period (50–77 °F). Lows in the single digits are not uncommon at night.

Low Humidity: Sambhar Lake has the dry climate typical of the desert all year round. The dry atmosphere helps the lake's surface water evaporate quickly, resulting in the creation of salt flats.

Dust Storms: Sambhar Lake, like the rest of Rajasthan, has dust storms and high winds in the summer. Visibility during dust storms can be severely reduced, making travel difficult.

Extreme Temperature Variations: There are large day-night and seasonal temperature swings at Sambhar Lake. There might be a big difference between the daytime and nighttime temperatures. The desert's extremes are on full display during the seasonal shift from hot summers to relatively cool winters.

8.5 Hydrology

Sambhar Lake in Rajasthan is endorheic, which means it does not have an outlet and its water does not flow out to the sea, both of which affect the lake's hydrology. It's hydrology may be broken down into the following basic components:

Catchment Area: The area from which water flows into the lake, known as its "catchment area", is rather modest. The monsoon rains bring small streams and hills into the area, making up the bulk of the landscape. Most of the land in the catchment region is desert or semiarid.

Rainfall: Sambhar Lake's proximity to the dry Thar Desert means that it sees very little precipitation. Between 200 and 400 mm of rain falls annually on average here (8–16 in.). Precipitation is heaviest during the monsoon season, which normally runs from July through September.

Water Sources: Sambhar Lake relies primarily on precipitation for its water supply. The lake gets refilled with rainwater from the catchment area every year during the monsoons. However, because of the low annual precipitation and high annual evaporation rates, the lake's water levels vary greatly.

Salinity: With a salt content between 8% and 10%, Sambhar Lake is extremely salty. Dry weather, limited amounts of rain, and evaporation from the lake's surface all contribute to its high salinity. Due to the increased concentration of the dissolved salts as water evaporates, salt flats are formed and salt is harvested.

Water Quality: Sambhar Lake's water is brackish, meaning it contains both fresh and salty elements. Most aquatic organisms cannot survive in the lake due to its brackish water and high salinity. Certain salt-loving creatures thrive in the lake, and many migrating birds stop there during the monsoons.

Fluctuating Water Levels: Sambhar Lake's water level fluctuates greatly since it receives no large incoming water and has no considerable outgoing water. The lake dries up during the dry season, exposing extensive salt plains. However, the lake's surface area can increase dramatically as water levels rise during the monsoon season (Kaushik and Raza 2019; Khandelwal 2022; Milja et al. 2011).

8.6 Limnology

Sambhar Lake in Rajasthan is the subject of limnology, which is the scientific study of lakes. Sambhar Lake is unlike any other lake in the world due to its extreme salinity and endorheic nature. The physico-chemical parameters are given in Table 8.2. The limnology of Sambhar Lake can be summed up as follows:

Salinity: With a salt content between 8% and 10%, Sambhar Lake is extremely salty. The high salinity is due to the dry climate, minimal precipitation, and high rates of evaporation. Most aquatic creatures cannot tolerate the high quantities of salt in the lake's brackish water, which is a mixture of freshwater and saltwater.

Water Chemistry: Sambhar Lake water is notable for its high mineral concentration. The lake water mostly contains sodium, chloride, sulphate, and magnesium ions. The carbonate and bicarbonate ions in the water give it a generally alkaline pH.

Nutrient Levels: The nutrition levels of Sambhar Lake are usually quite low. Low nutrient availability and excessive salinity make the water unsuitable for most types of algae and aquatic plants. There may be an influx of nutrients from the surrounding catchment region, however, when water levels rise as they do during the monsoon season.

Seasonal Fluctuations: The limnological parameters of Sambhar Lake change dramatically throughout the year. The lake's salty content rises throughout the dry season due to the increased concentration of dissolved salts when the water evaporates. As the oceans dry up, they leave behind vast salt plains. While the lake's salinity remains relatively constant year-round, its water levels rise during the monsoon season.

Biotic Communities: Sambhar Lake is home to a few halophilic (salt-loving) microorganisms and invertebrates despite its low biodiversity due to the lake's high salinity. Some bacteria, algae, and crustaceans have adapted to survive in these extremely salty environments. Birds that migrate south for the winter find refuge in the lake's brackish water and abundant food supply.

Environmental Challenges: Water pollution from human activities and shifting land use patterns are two examples of the environmental threats that threaten Sambhar Lake. Lake water levels and salinity can be affected by the overexploitation of groundwater for agricultural and industrial uses.

Table 8.2 List of physico-chemical parameters and their importance

S. no.	Parameter	Importance for lake	Importance for salt production
1	Water temperature	Ecological balance	Affects salt crystallisation and harvest
2	pH (acidity/alkalinity)	Aquatic life health	Influences salt quality and yield
3	Dissolved oxygen (DO)	Vital for fish	Essential for brine quality and salt crystallisation
4	Turbidity	Water clarity	Can impact salt harvesting and brine quality
5	Salinity	Fundamental for ecology	High salinity crucial for salt production
6	Algal blooms	Ecological health	Can interfere with salt harvesting and quality
7	Nitrate concentration	Nutrient management	Nitrate can affect salt quality and the environment
8	Phosphate concentration	Nutrient control	Phosphate levels influence salt production efficiency
9	Total suspended solids (TSS)	Water quality	Affects salt crystallisation and brine quality
10	Calcium concentration	Aquatic organisms	Influences salt precipitation and purity
11	Magnesium concentration	Water hardness	Affects salt quality and can lead to scaling issues
12	Potassium concentration	Nutrient management	Impacts salt composition and quality
13	Sulphate concentration	Water chemistry	Sulphate can affect brine quality and salt purity
14	Chloride concentration	Salinity indicator	High chloride is essential for salt production
15	Total dissolved solids (TDS)	Water quality	TDS levels are closely tied to salt production
16	Sedimentation rate	Erosion control	Influences sediment removal and salt harvesting
17	Wind speed and direction	Affects water mixing	Can impact salt evaporation and crystal formation
18	Rainfall	Water input	Rainfall can dilute brine concentration in the lake
19	Solar radiation	Evaporation rates	Solar radiation drives evaporation and salt production
20	Nutrient concentrations (e.g., N, P, K)	Ecological health	Nutrient levels can influence algae and salt yield

8.7 Biodiversity

Despite its high salinity, Rajasthan's Sambhar Lake is home to a wide variety of plant and animal life. Despite its high salinity, Rajasthan's Sambhar Lake is home to a wide variety of plant and animal life which is given in Fig. 8.3 and Table 8.3 about fauna and Table 8.4 about flora. Sambhar Lake is home to a wide variety of organisms, including but not limited to:

Avian Species: The variety of species found at Sambhar Lake is well-known, and the lake is especially important as a wintering ground for migratory birds. A wide variety of waterbirds, such as flamingos, pelicans, storks, ducks, geese, waders, and shorebirds, congregate around the lake. It is a major birding hotspot, with over 200 different bird species having been spotted there. Some of the migratory birds are given in Fig. 8.4.

Flamingos: The flamingos at the lake are what really draw people there. Sambhar Lake is home to two species of flamingo: the Greater (*Phoenicopterus roseus*) and the Lesser (*Phoeniconaias minor*). The sight of these beautiful birds feeding and nesting in the shallow waters of the lake is captivating.

Resident Birds: Numerous resident bird species can be seen around Sambhar Lake, in addition to the migratory ones. Heron, egret, cormorant, ibis, tern, gull, and raptor species like eagles and vultures are all examples. These native bird species make their homes in the lake's marshy sections and the adjacent marshes.

Other Fauna: Even though birds make up the bulk of Sambhar Lake's biodiversity, the lake is home to a wide variety of other organisms as well. These include brine shrimp and other crustaceans that have adapted to the salty environment. In addition, reptiles, animals, and desert-adapted plant species thrive in the dry landscape that surrounds the lake.

Conservation Significance: Due to its high bird population, Sambhar Lake has been designated as an Important Bird Area (IBA) by BirdLife International. Due to the ecological significance of the lake's unique birdlife, conservation efforts have been made to maintain and manage the lake and its adjacent wetlands (Naik and Sharma 2021; Naik and Sharma 2022a, b; Rajashree et al. 2023).

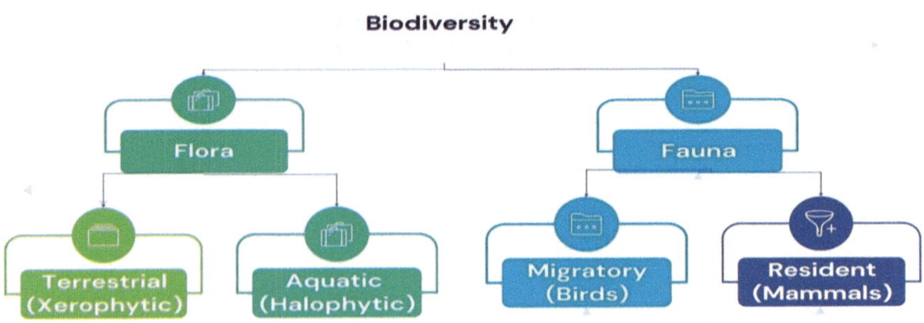

Fig. 8.3 Biodiversity of Sambhar Lake

Table 8.3 Interdependence of lake and fauna of Sambhar Lake

S. no.	Fauna	Habitat requirement	Benefit to Sambhar Salt Lake
1	Flamingos (greater and lesser)	Shallow brackish water, mudflats	Ecotourism, algae control, nutrient cycling
2	Plovers and sandpipers	Sandy and muddy shores	Shoreline protection, insect control
3	Gulls and terns	Open waters and sandbanks	Scavengers, fish population control
4	Coots	Reed beds, open water	Aquatic plant control, nutrient cycling
5	Ducks (shoveler, pintail)	Open water, submerged vegetation	Wetland habitat, aquatic plant control
6	Waders (stilts, avocets)	Mudflats, shallow water	Insect control, nutrient cycling
7	Crustaceans (artemia, brine shrimp)	Brine pools, hypersaline waters	Food source for waterfowl, nutrient cycling
8	Insects (midges, mosquitoes)	Brine pools, mudflats, reed beds	Part of the food chain for bird species
9	Fish (mullet, catfish)	Brackish and hypersaline waters	Food source for birds, nutrient cycling
10	Microorganisms (halophiles)	Brine pools, salt crusts	Salt production, microbial ecology
11	Algae (*Dunaliella*, diatoms)	Brine pools, open waters	Source of food for Artemia and birds
12	Marsh plants (reeds, phragmites)	Marshes and shallow water	Shoreline stabilisation, habitat for fauna
13	Tadpoles and frogs	Temporary water bodies, marshes	Aquatic food chain, insect control
14	Turtles (terrapins, softshells)	Shallow waters, mud banks	Ecosystem balance, insect control
15	Snakes (rat snakes, water snakes)	Marshes, reed beds, mudflats	Natural predators, balance in food chain
16	Molluscs (snails, clams)	Shallow water, muddy substrate	Nutrient cycling, filter-feeders
17	Cranes (*sarus, demoiselle*)	Open waters, grassy areas	Ecotourism, wetland preservation
18	Water buffalo	Marshes, muddy areas, reed beds	Vegetation control, nutrient cycling
19	Jackals	Surrounding dry areas	Scavengers, part of the ecosystem

Table 8.4 Interdependence of flora and the Sambhar Lake

S. no.	Flora	Habitat requirement	Benefit to Sambhar Salt Lake
1	Saltwort (*Salsola*)	Saline and alkaline soils	Stabilises soil, salt production
2	Seashore Paspalum (*Paspalum vaginatum*)	Saline soils, marshes	Erosion control provides bird habitat
3	Common reed (*Phragmites australis*)	Marshes, shallow water	Habitat for birds, shoreline stability
4	Red mangrove (*Rhizophora mucronata*)	Brackish water, mudflats	Coastal protection, fish nursery
5	Algal mats (various species)	Brine pools, open waters	Primary producer, food for fauna
6	Halophytes (various species)	Salt-tolerant plants	Erosion control, habitat for invertebrates
7	Bulrushes (*Schoenoplectus* spp.)	Marshes, wetlands	Habitat for waterfowl, shoreline protection
8	Water hyacinth (*Eichhornia crassipes*)	Freshwater bodies, slow-moving waters	Aquatic plant control, habitat for insects
9	Pondweed (*Potamogeton* spp.)	Freshwater, submerged in lakes	Aquatic plant control improves water quality
10	Waterlilies (*Nymphaea* spp.)	Calm, shallow waters	Aesthetic value provides shade and habitat
11	Cattails (*Typha* spp.)	Marshes, wetlands	Habitat for birds, nutrient cycling
12	Spikerushes (*Eleocharis* spp.)	Shallow waters, wetlands	Erosion control, habitat for aquatic fauna
13	Tamarisk (*Tamarix* spp.)	Saline and alkaline soils	Erosion control stabilises soil
14	Alkali sacaton (*Sporobolus airoides*)	Alkaline soils, grasslands	Soil stabilisation, food for herbivores
15	Indian wild rice (*Zizania latifolia*)	Shallow freshwater, marshes	Habitat for birds, nutrient cycling
16	Salicornia (glasswort)	Saline soils, marshes	Nutrient cycling, habitat for invertebrates
17	Desert marigold (*Baileya pleniradiata*)	Arid and alkaline soils	Erosion control, salt crust stabilisation
18	Prosopis (*Mesquite*)	Sandy soils, arid conditions	Windbreak, habitat for wildlife
19	Lantana camara (common lantana)	Dry and open areas	Erosion control attracts pollinators
20	Indian Frankincense (*Boswellia serrata*)	Dry and rocky terrain	Soil conservation, potential for resin

Fig. 8.4 Migratory birds of Sambhar Lake

8.8 Opportunities of Livelihood

BirdLife International has designated Sambhar Lake as an IBA because of the high number of birds that live there. Conservation efforts have been made to maintain and manage the lake and the adjacent wetlands because of the ecological significance of the lake's rich wildlife. The livelihood options of the villagers are given in Figs. 8.5 and 8.6 and economic importance of biodiversity in Table 8.5.

Salt Production: One of India's primary salt producers is Sambhar Lake. Salt can be easily extracted from the lake because of its high salinity and surrounding salt flats. Numerous salt factories give locals a chance to make a living and contribute to the economy. Work opportunities in the salt industry range from mining to processing to packing to shipping.

Tourism and Hospitality: Sambhar Lake is a popular destination for sightseers, birdwatchers, and ecotourists due to its picturesque terrain and abundant biodiversity. Hotels, inns, resorts, restaurants, and tour operators are just some of the industries that stand to benefit from this trend. To further highlight the lake's avian diversity, boat rides, nature walks, and birdwatching tours can all be arranged with a knowledgeable guide.

Handicrafts and Souvenirs: Souvenir and craft shops can thrive thanks to the area's rich cultural history. Traditional Rajasthani handicrafts like pottery, textiles, woodwork crafts, and jewellery can be made and sold by local artists. Tourists

Fig. 8.5 Livelihood options at Sambhar Lake

Fig. 8.6 Livelihood of villagers (**a**–**e** are primary activities and **f**, **g** are secondary activities)

Table 8.5 Ecological and economic importance of biodiversity of Sambhar Lake

S. no.	Flora/fauna	Ecological importance	Economic importance
1	Flamingos	Biodiversity, nutrient cycling, tourism attraction	Ecotourism, bird watching, potential for local businesses
2	Gulls and terns	Predators of aquatic organisms contribute to nutrient cycling	Tourist attractions, bird watching opportunities
3	Artemia (brine shrimp)	Vital link in the food chain maintains brine quality	Harvested for fish feed and aquaculture industry
4	Seashore Paspalum	Erosion control, habitat for bird species	Supports shoreline stabilisation and tourism
5	Halophytes (salt-tolerant plants)	Soil stabilisation, ecological diversity	Traditional uses in local medicine, fuel, and fodder
6	Microorganisms (halophiles)	Salt production, nutrient cycling, microbial ecology	Involved in the salt-making process
7	Coots	Vegetation control, part of the aquatic food chain	Potential for bird-watching tourism
8	Mudskippers	Adapted to low oxygen environments, part of the food chain	Potential for ecotourism and research interest
9	Tadpoles and frogs	Part of the aquatic food web, insect control	Limited economic significance, but contributes to ecosystem balance
10	Brine flies	Part of the food chain serves as food for birds	Not a significant economic resource, but supports the ecosystem
11	Indian wild rice	Habitat for bird species, nutrient cycling	Traditional uses for food and fodder, minor economic value
12	Molluscs (snails, clams)	Filter feeders, nutrient cycling, part of the food chain	Limited economic value, but supports the lake's ecosystem
13	Jackals	Scavengers, part of the food web, maintain balance	Not a direct economic resource, but plays a role in the ecosystem
14	Algal mats	Primary producers support food web	Limited economic value, but vital for lake ecology
15	Shovelers and pintail ducks	Insect control, part of the aquatic food chain	Potential for bird-watching tourism and conservation
16	Water buffalo	Vegetation control, wetland management	Used in salt production and for transportation in the region
17	Saltwort	Soil stabilisation, part of the food chain	Traditional uses as food and medicine, minor economic value
18	Lantana camara (common lantana)	Erosion control, pollinator attraction	Not a direct economic resource, but supports the ecosystem
19	Sand dunes	Influence landscape and water flow	Limited economic importance, part of the local environment
20	Indian Frankincense	Soil conservation, potential for resin extraction	Traditional use for incense and medicine

wishing to purchase genuine, regionally produced mementoes can get what they need here.

Agriculture and Farming: The surrounding areas of Sambhar Lake, though arid, support some agriculture. Farmers cultivate crops such as wheat, mustard, barley, and pulses using irrigation techniques. Additionally, animal husbandry, including dairy farming and rearing livestock like goats and sheep, can provide livelihood opportunities. As a result of its rich history and customs, the area is ripe for the souvenir and craft industries. Craftsmen in Rajasthan can make and market the region's signature jewels, ceramics, textiles, and wooden works. Shoppers who want genuine, regionally produced trinkets can find what they need here.

Fishing: Sambhar Lake is home to certain traditional fishing practices, albeit they are severely restricted by the lake's extreme salinity. Small-scale fishing operations run by locals focus on brackish-water-adapted species. Those who catch fish can sell their catch at local or regional marketplaces.

Eco-Tourism and Conservation: Sambhar Lake's diverse bird population and otherworldly habitat present excellent prospects for ecotourism and conservation efforts. To preserve the lake's ecosystem, it may be necessary to establish sustainable practices, perform research on birdlife and wetland protection, and organise educational initiatives.

Transportation and Services: Taxis, auto-rickshaws, and bike rentals are all needed to accommodate the growing number of tourists visiting Sambhar Lake. Other areas where you can find success include tour guiding, photography, event planning, and operating a store that sells necessities (Sahay et al. 2012; Sharma et al. 2022a, b; Shukal and Rahaman 2006).

8.9 Opportunities of Tourism

The saline atmosphere, avian diversity, and cultural legacy of Rajasthan's Sambhar Lake make it a prime tourist destination. The opportunity for bird tourism along with academic and research scope is shown in Fig. 8.7. The tourism industry in Sambhar Lake can benefit greatly from the following:

Birdwatching and Wildlife Tourism: Sambhar Lake is a haven for those who enjoy outdoors and watching birds. Flamingos, pelicans, storks, and many more waterbirds are just some of the migratory birds that stop by the lake every year. Tourists interested in seeing and documenting the avian species can be attracted by offering birding tours, guided walks, and photography opportunities.

Nature Photography: Sambhar Lake's beautiful scenery, expansive salt flats, and abundant fauna make it a prime location for wildlife photography. Many photographers travel to the lake in search of the perfect shot of the setting sun, a flock of birds, or the lake's other-worldly landscape. This subset of the market can be served by companies offering photography courses, excursions, and other such activities.

Cultural and Heritage Tourism: Sambhar Lake and its surrounding area are steeped in Rajasthani history and tradition. Villages and towns in the area provide old

Fig. 8.7 Opportunities of academic and research in the lake (**a**, **b** show academic visit; **c**, **d** show research and bird census by Central University of Rajasthan and **e**, **f** shows amusement)

temples, traditional festivals, and traditional arts for visitors to experience. A deeper understanding of Rajasthan's thriving traditions and way of life can be gained by promoting cultural excursions, history walks, and village visits.

Adventure Tourism: Sambhar Lake's expansiveness and desolate surroundings make it a prime location for extreme travel. Desert camping, camel safaris, off-road biking, and other adventure activities can keep visitors coming back for more. Tourists seeking exhilarating experiences can find them here, thanks to the area's distinctive landscape and breathtaking vistas. Sambhar Lake's expansiveness and desolate surroundings make it a prime location for extreme travel. Desert camping, camel safaris, off-road biking, and other adventure activities can keep visitors coming back for more. Tourists seeking exhilarating experiences can find them here, thanks to the area's distinctive landscape and breathtaking vistas.

Wellness and Relaxation: Sambhar Lake's tranquil atmosphere and beautiful scenery make it a great spot for health and wellness vacations. Spas, yoga studios, and other relaxation establishments that take advantage of the lake's peaceful setting would be popular with vacationers looking to unwind.

Cultural Festivals: Sambhar Lake is the site of many celebrations and activities that honour local culture and history. Folk music, dancing performances, local handicrafts, and traditional cuisine are all on display at festivals like the Sambhar Mahotsav. Tourists looking for an authentic local experience can be attracted by organising and advertising for such gatherings.

Educational and Research Tourism: Sambhar Lake's rare habitat is a draw for academic and scientific travellers. Field studies, research initiatives, and instructional programmes focusing on avian biodiversity, wetland ecology, and environmental conservation can all benefit from a visit to the lake from academic institutions, research groups, and wildlife enthusiasts (Singh et al. 2013; Singh et al. 2019; Sinha 2014; Sinha and Raymahashay 2004).

8.10 Threats and Challenges

The ecological vitality, biodiversity, and prospects for sustainable development of Rajasthan's Sambhar Lake are threatened by a number of factors. Figure 8.8 shows threats of the lake. The main dangers and difficulties that Sambhar Lake has to deal with are as follows:

Fig. 8.8 Threats to Sambhar Lake (**a** shows dry lake near Shakambari temple; **b** shows dumping of waste near lake periphery, **c** shows construction within the lake, **d** shows illegal groundwater pumping, **e** shows private construction near lake, and **f** shows avian botulism impact)

Water Pollution: Human activity in the region poses a threat to the lake's water quality. Lake water can be tainted by various sources, including industrial discharge, agricultural runoff including fertilisers and pesticides, and household trash. The lake's environment, vegetation, and wildlife may all be negatively affected by pollution, and the water quality may deteriorate as a result.

Habitat Loss and Fragmentation: Loss of habitat and fragmentation are potential consequences of rapid urbanisation, encroachments, and infrastructural development in the Sambhar Lake area. Destroying or degrading wetland regions and natural wildlife habitats can reduce the amount of space and resources available to species that rely on the lake.

Overexploitation of Natural Resources: Sambhar Lake's water level may drop if groundwater is pumped out for agricultural and industrial use. Withdrawing too much water from a lake can disrupt its hydrology, raise salinity levels, and throw off the ecosystem's delicate balance. The lake's environment and local people's way of life are vulnerable to the negative effects of overexploitation of resources like salt.

Climate Change and Drought: Sambhar Lake faces a serious threat from climate change. Water supply and lake salinity are both affected by climate change factors such as rising temperatures, shifting precipitation patterns, and extended droughts. The breeding and feeding cycles of migrating birds can be thrown off by shifts in the weather.

Invasive Species: The lake's surroundings and marshes are vulnerable to invasion by invasive plant species such as *Prosopis juliflora* (mesquite). These invasive species impair habitat quality for wildlife by outcompeting native flora and throwing the ecosystem out of whack. Invasive species must be managed and controlled in order to protect the lake's ecosystem.

Lack of Awareness and Conservation Efforts: Limited awareness about the ecological importance and conservation needs of Sambhar Lake can hinder efforts to protect and preserve the lake. Educational programs, community engagement, and awareness campaigns are vital to promote sustainable practices, responsible tourism, and the importance of preserving the lake's biodiversity. *Prosopis juliflora* (mesquite), an invasive plant, has the potential to spread into the wetland areas and landscape around the lake. These invasive species decrease the quality of animal habitat by outcompeting native plant life. The biological health of the lake depends on effective control and prevention of invading species (Srikantia 2009; Sundaresan et al. 2006; Vijay et al. 2018; Vijay et al. 2016) (Table 8.6).

Various groups, including governments, local communities, conservation organisations, and tourists, must work together to face these dangers and problems. To protect the ecological and cultural significance of Sambhar Lake, it is essential to employ sustainable development techniques, enforce rules to limit pollution, encourage eco-tourism, and launch conservation programmes.

Table 8.6 Sources of threats and the solutions to the threats of Sambhar Lake

S. no.	Threat	Causes of threats	Solutions to deal with the threat
1	Water pollution	Industrial discharge, agricultural runoff, urban sewage	Implement water quality regulations and treatment measures, promote responsible waste disposal practices
2	Habitat destruction	Urban development, infrastructure projects, salt production activities	Establish protected areas, habitat restoration, and responsible land use planning
3	Invasive species	Introduction of non-native species, trade, and human activities	Monitoring and management of invasive species, quarantine measures
4	Overfishing	Excessive fishing, unsustainable practices	Implement and enforce fishing regulations, promote sustainable fishing methods
5	Climate change	Rising temperatures, altered precipitation patterns	Adaptation measures, promote carbon reduction, and climate-resilient ecosystem management
6	Water scarcity	Reduced freshwater inflow, increased water demand	Water conservation, sustainable water use practices, promote efficient irrigation methods
7	Soil erosion	Deforestation, land degradation, and land use practices	Soil conservation measures, reforestation, and erosion control techniques
8	Pollution from salt production	Brine discharge, chemical usage	Improved salt production practices, responsible brine disposal, and pollution monitoring
9	Bird disturbance	Human activities, tourism, and habitat disruption	Implement bird-watching guidelines, regulate tourism activities, and establish bird sanctuaries
10	Altered hydrology	Infrastructure projects, water diversion, and changing hydrological patterns	Evaluate and manage hydrological alterations, consider environmental impacts in development projects

References

Bhat AH, Sharma KC, Banday UJ (2015) Impact of climatic variability on salt production in Sambhar Lake, a Ramsar wetland of Rajasthan, India. Middle-East J Sci Res 23(9):2060–2065

British Library (n.d.). https://www.gandhisaltmarch90.com/exhibition-archive/sambhar-lake-01. Assessed 30 Oct 2023

Cherekar MN, Pathak AP (2015) Studies on haloalkaliphilic gammaproteobacteria from hypersaline Sambhar Lake, Rajasthan, India. Indian J Geomar Sci 44(10):1646–1653

Gupta M, Aggarwal S, Navani NK, Choudhury B (2015) Isolation and characterization of a protease-producing novel haloalkaliphilic bacterium Halobiforma sp. strain BNMIITR from Sambhar Lake in Rajasthan, India. Ann Microbiol 65(2):677–686

Jain AK, Das SK, Goyal SA (2005) Conservation planning of Sambhar Lake, Rajasthan using satellite remote sensing and GIS. A thesis for M. Tech, Indian Institute of Remote Sensing National Remote Sensing Agency, Department of Space, Government of India. Submitted to Andhra University, Andhra Pradesh, India

Jakher GR, Bhargava SC, Sinha RK (1990) Comparative limnology of Sambhar and Didwana lakes (Rajasthan, NW India). In: Saline Lakes: proceedings of the fourth international sym-

posium on Athalassic (inland) Saline Lakes, held at Banyoles, Spain, May 1988. Springer Netherlands, Basel, pp 245–256
Kaushik G, Raza K (2019) Potential of novel Dunaliella salina from Sambhar Salt Lake, India, for bioremediation of hexavalent chromium from aqueous effluents: an optimized green approach. Ecotoxicol Environ Saf 180:430–438
Khandelwal S (2022) Impact of land use land cover change on Sambhar Lake. In: Addressing environmental challenges through spatial planning. IGI Global, Hershey, pp 127–151
Milja TE, Prathish KP, Rao TP (2011) Synthesis of surface imprinted nanospheres for selective removal of uranium from simulants of Sambhar Salt Lake and ground water. J Hazard Mater 188(1–3):384–390
Naik R, Sharma L (2021) Spatio-temporal modelling for the evaluation of an altered Indian saline Ramsar site and its drivers for ecosystem management and restoration. PLoS One 16(7):e0248543
Naik R, Sharma LK (2022a) Monitoring migratory birds of India's largest shallow saline Ramsar site (Sambhar Lake) using geospatial data for wetland restoration. Wetl Ecol Manag 30(3):477–496
Naik R, Sharma LK (2022b) Sentinel-2 red-edge band as virus indicators for sustainable wetland management assessed during COVID-19 lockdown. In: GIScience for the sustainable management of water resources. Apple Academic Press, Oakville, pp 165–183
Rajashree N, Laxmi Kant S, Avinash S (2023) Characterization of saline soil for the halophytes of largest inland saline wetland of India using geospatial technology. J Oceanol Limnol 41:1–15
Sahay H, Mahfooz S, Singh AK, Singh S, Kaushik R, Saxena AK, Arora DK (2012) Exploration and characterization of agriculturally and industrially important haloalkaliphilic bacteria from environmental samples of hypersaline Sambhar Lake, India. World J Microbiol Biotechnol 28:3207–3217
Sharma L, Naik R, Raj A (2022a) Inland saline wetlandscapes: the missing links for 4th Ramsar strategic plan (2016-2024) in India. In: Research anthology on measuring and achieving sustainable development goals. IGI Global, Hershey, pp 1062–1083
Sharma P, Rekhi P, Kumari S, Debnath M (2022b) Deciphering the molecular diversity of related halophilic Bacillus sp. isolated from Sambhar Lake and the functional characterizations of surfactin. J Surfactant Deterg 25(6):757–774
Shukal DW, Rahaman AA (2006) Sambhar Lake a wetland—an assessment. In: Proceedings of the 1st international conference on the ecological importance of solar saltworks (CEISSA 06), Santorini Island, Greece
Singh BP, Neha S, Singh SP (2013) Modern salt (halite) deposits of the Sambhar Lake, Rajasthan and their formative conditions. Curr Sci 104(11):1482–1484
Singh P, Khadim R, Singh AK, Singh U, Maurya P, Tiwari A, Asthana RK (2019) Biochemical and physiological characterization of a halotolerant Dunaliella salina isolated from hypersaline Sambhar Lake, India. J Phycol 55(1):60–73
Sinha R (2014) The Sambhar Lake: the largest saline Lake in northwestern India. In: Landscapes and landforms of India. Springer, Berlin, pp 239–244
Sinha R, Raymahashay BC (2004) Evaporite mineralogy and geochemical evolution of the Sambhar Salt Lake, Rajasthan, India. Sediment Geol 166(1–2):59–71
Srikantia SV (2009) In: Das BK (ed) Lakes: water and sediment geochemistry, vol 74. Sathish Serial Publishing House, New Delhi, p 275
Sundaresan S, Ponnuchamy K, Rahaman AA (2006) Biological management of Sambhar Lake saltworks (Rajasthan, India). In: Proceedings of the 1st international conference on the ecological importance of solar saltworks, 20–22 October 2006, Santorini, Greece, pp 199–208
Vijay R, Pinto SM, Kushwaha VK, Pal S, Nandy T (2016) A multi-temporal analysis for change assessment and estimation of algal bloom in Sambhar Lake, Rajasthan, India. Environ Monit Assess 188:1–10
Vijay A, Arora S, Gupta S, Chhabra M (2018) Halophilic starch degrading bacteria isolated from Sambhar Lake, India, as potential anode catalyst in microbial fuel cell: a promising process for saline water treatment. Bioresour Technol 256:391–398

Part IV

Wetlands Restoration and Conservation

Need for Wetland Restoration 9

Abstract

Wetlands, renowned for their productivity, confront formidable threats from human activities and climate change. This section accentuates the pivotal role of wetland restoration, which not only enhances water quality and biodiversity but also fortifies resilience against natural calamities. The imperative for restoration is underscored by wetlands' significant contribution to climate change mitigation through carbon sequestration and greenhouse gas regulation. Saline wetlands, distinctive ecosystems prevalent in arid and semi-arid regions with high salt concentrations, demand urgent restoration efforts. These ecosystems offer vital habitats for specialized flora and fauna while acting as natural barriers against saltwater intrusion. Nevertheless, restoring saline wetlands poses distinctive challenges due to elevated salinity levels, adverse soil conditions, and harsh climates. Overcoming these obstacles necessitates specialized knowledge and innovative technologies tailored to the intricacies of saline wetland restoration, ensuring the preservation of their delicate equilibrium.

Keywords

Environmental quality improvement · Natural resilience · Climatic disasters

9.1 Applicability of Technologies for Restoration

The term "ecosystem restoration" describes the procedure of bringing back an ecosystem's original or enhanced ecological performance after it has been degraded or destroyed. Restoring an ecosystem entails many different steps taken to improve its biodiversity, natural processes, and ecological services. Forests, wetlands, rivers, coral reefs, and even urban areas are all potential sites for ecosystem restoration.

L. K. Sharma, R. Naik, *Conservation of Saline Wetland Ecosystems*, https://doi.org/10.1007/978-981-97-5069-6_9

Important components of ecological restoration include:

Assessing the Ecosystem: Assessing the ecosystem's current health and identifying the causes and amount of deterioration is the first stage in restoring an ecosystem. Understanding the historical dynamics of an ecosystem requires research into its ecological, hydrological, and biological aspects as well as the causes of its decline.

Setting Restoration Goals: The assessment serves as the basis for developing restoration targets and aims. Restoring biodiversity, bettering water quality, increasing ecosystem services, protecting endangered species, and recreating specific habitat types are all examples of such goals. To direct the restoration process, precise and quantifiable goals are set at the outset.

Planning and Implementation: A comprehensive restoration strategy and set of procedures are established. Reforestation, habitat building, invasive species removal, reintroducing native species, preventing soil erosion, and restoring water flow are all possible components of this approach. Planting, building, and other forms of restoration work are examples of the kinds of operations that happen on the ground during the implementation phase.

Monitoring and Adaptive Management: It is essential to keep tabs on how things are going with the restoration work. The success of restoration efforts can be gauged, ecosystem recovery assessed, and course corrections made with the aid of regular monitoring. With adaptive management in place, restoration plans can be adjusted in light of new data and shifting circumstances.

Stakeholder Engagement: Communities, landowners, government agencies, NGOs, and scientists are all important players in ecosystem restoration that must work together. Involving these parties at every stage of the restoration process increases their investment in the project, solidifies public support, and incorporates their insights and expertise.

Long-term Management and Maintenance: Restoring an ecosystem is an ongoing process. The longevity and resilience of the restored ecosystem are dependent on regular management and maintenance efforts. Invasive species eradication, fire prevention, habitat observation, and public involvement in conservation and restoration efforts are all possible means to this end.

Environmental deterioration, biodiversity loss, and the loss of ecosystem services that humans and other organisms rely on are just a few of the many reasons why ecosystem restoration is so important. Ecosystem restoration can improve ecosystems' ability to supply clean water, store carbon, provide for wildlife habitats, moderate climate, and contribute to the planet's health and well-being (Allison 2023; Clewell and Aronson 2012).

9.1.1 Degraded Status of Wetlands

Habitat Loss and Fragmentation: For the sake of agriculture, urban expansion, infrastructure projects, and aquaculture, wetland areas are frequently drained, filled, or converted. This causes wetland habitats to be lost or fragmented, breaking up biological connections and decreasing the amount of land available to species that rely on wetlands for survival.

Pollution: Pollution from factories, farms, and homes can have devastating effects on wetland ecosystems. Excessive nutrient inputs, sewage and industrial effluent discharge, and runoff from agricultural fields are the primary causes of water pollution and eutrophication. Wetland vegetation, aquatic life, and ecological harmony can all suffer from pollution.

Invasive Species: Wetlands can be colonised by invasive plant and animal species, which can change the dynamics of the ecosystem by outcompeting native species. Dense mats of invasive species like water hyacinth and purple loosestrife can block streams, block out sunlight, and drive out native plant and animal life.

Overexploitation of Resources: Degradation of wetland ecosystems can result from the excessive use of wetland resources such as water, fish, timber, and non-timber forest products. The loss of biodiversity, the disruption of food chains, and the modification of hydrological patterns can result from unsustainable fishing activities, excessive water extraction, and illegal logging.

Climate Change: The effects of climate change, such as increased temperatures, changing precipitation patterns, and sea level rise, pose a serious threat to wetland ecosystems. Species ranges fluctuate, and unique wetland habitats may be lost, as a result of these alterations to wetland hydrology, salinity levels, and vegetation composition.

Drainage and Alteration of Hydrology: Draining a wetland for agricultural or urban development objectives might change the water table and disrupt natural hydrological processes. Wetland vegetation, hydrological processes, and ecological services as a whole are all susceptible to being disrupted by human activities.

Unsustainable Practices: Wetland ecosystems can be degraded by unsustainable land and water management practices such as excessive water extraction, overgrazing, and improper waste management. Soil erosion, water scarcity, decreased water quality, and biodiversity loss are all possible outcomes of such practices.

Degradation of wetland areas has serious effects on ecosystems and human health. Critical ecosystem services provided by wetlands include water purification, flood control, climate control, carbon sequestration, and the provision of habitat. Effective conservation and restoration methods, sustainable land and water management practice, and strong policy frameworks that prioritise the protection and sustainable use of wetland ecosystems are required to address the degraded status of wetlands.

Biodiversity Conservation: Wetlands host a huge variety of plant and animal species, including many that are currently at risk of extinction, making them one of the most biologically varied environments on Earth. Restoring degraded wetland areas helps maintain and increase biodiversity by offering new homes for plants and animals.

Climate Change Mitigation: The importance of wetland ecosystems in slowing global warming is crucial. They help lower emissions of greenhouse gases because of the massive amounts of carbon they store and sequester. Wetlands that have been restored increase their capacity to trap and store carbon, aiding in the fight against global warming.

Water Quality Improvement: Wetlands remove sediments and impurities from water through natural filtration. Restoring wetland areas helps clean up water supplies by decreasing runoff into nearby bodies of water. The supply of clean water for human consumption and the preservation of downstream ecosystems depend on this.

Flood Control and Water Regulation: During times of extreme precipitation or flooding, wetland areas can be used as a "natural sponge", soaking up and storing the surplus water. Wetlands can be restored to help reduce the danger of downstream floods by increasing storage capacity and lowering water velocity. This is of paramount importance in flood-prone regions or wherever human populations or critical infrastructure are at risk.

Erosion Control and Sediment Retention: Wetlands slow the flow of water and trap silt, both of which contribute to soil stabilisation and the prevention of erosion. In order to keep healthy soil profiles and prevent the loss of fertile land, restoring wetlands can limit soil erosion, protect coasts from erosion, and retain sediments.

Sustainable Livelihoods: Wetland habitats support the economies of many places through activities including fishing, farming, tourism, and customs. By enhancing fisheries, expanding agricultural prospects, and drawing eco-tourists, wetland restoration helps preserve local economies and protects cultural traditions.

Resilience to Climate Change and Natural Disasters: Wetlands protect against the effects of climate change and natural disasters by acting as a buffer zone. They can mitigate the effects of storm surges, provide a safe haven for species during drought or other environmental stresses, and serve as a habitat in general. The overall resilience of ecosystems and populations can be strengthened by restoring wetlands.

Government agencies, conservation groups, local communities, and other interested parties must work together to restore wetlands. Sustainable land and water management practices are implemented, as well as the restoration of hydrological linkages, the eradication of invasive species, the return of native vegetation, and so on. It is possible to promote ecosystem health, maintain biodiversity, reduce the effects of climate change, and guarantee the long-term preservation of these priceless ecosystems by recovering degraded wetland areas (Comín 2010; Higgs 2003; Holl 2020; Kent 2000).

9.1.2 Need for Saline Wetland Restoration

Unique Ecosystems: Communities of plants and animals living in saline wetlands are highly specialised because of the salty environment. Salt-tolerant plant species, migratory bird populations, and an abundance of aquatic life are all supported by these environments. Saline wetland restoration aids in the conservation of these vital habitats for wildlife.

Habitat for Endangered Species: Several threatened species rely on saline wetland habitats because of their importance in supporting their unique survival strategies. Wetlands along the coast, for instance, play a crucial role in the survival of many migrating bird species. Saline wetland restoration can help save these species by shoring up and improving the ecosystem in which they exist.

Carbon Sequestration: Mangrove forests, salt marshes, and seagrass beds are all examples of saline wetlands, and they all serve as effective carbon sinks. They remove a lot of CO_2 from the air and store it in their vegetation and soils. By increasing their potential to store carbon, restoration of salt marshes aids in the fight against global warming.

Erosion Control and Coastal Protection: When it comes to safeguarding coastlines from storms and rising seas, saline wetlands are crucial. Stabilizing sediments and reducing wave erosion, mangroves and salt marsh grasses have extensive root systems. Maintaining coastal resilience, protecting coastal populations, and preserving rich coastal ecosystems can all be aided by restoring salty wetlands.

Water Quality Improvement: Saline wetlands are effective natural filters that mitigate the negative effects of fertiliser runoff and pollution on water supplies. Improve water quality and sustain the vitality of neighbouring marine ecosystems by allowing mangroves and salt marshes to take in and store pollutants, sediment, and excess nutrients. Saline wetland restoration can help improve coastal water quality.

Sustainable Livelihoods: Saline wetlands are important to many industries and ways of life, including the fishing, aquaculture, tourist, and salt production industries. The availability of fishing resources, the promotion of eco-tourism, and the support of sustainable salt production practices can all be aided by the restoration and conservation of saline wetland ecosystems.

Resilience to Climate Change: The effects of climate change, such as rising sea levels, storms, and coastal floods, can be mitigated by saline wetland ecosystems. Mangroves and salt marshes, with their intricate root systems and ability to absorb wave energy, provide a protective buffer for coastal settlements. The overall resilience of coastal areas is improved by restoring salty wetlands to their former state.

Water systems may be repaired, native plants replanted, invasive species eradicated, and sustainable management techniques put into place in an effort to revive salt marshes. The restoration of salty wetlands and the full realisation of the many advantages they give requires close cooperation between a wide range of interested parties, including local communities, government agencies, and conservation organisations.

9.1.3 Current Challenges for Saline Wetland Restoration

There are a number of obstacles that can reduce the success of restoration efforts in saline wetlands which are given in Table 9.1. Examples of such difficulties are:

Lack of Awareness and Understanding: Saline wetland restoration might be hampered by a lack of public appreciation for the ecosystem's worth and signifi-

Table 9.1 Challenges of saline wetlands and their ways of restoration

S. no.	Parameter of saline wetland	Causes of threats	Solutions for restoration
1	Salinity Levels	Natural factors, pollution, climate change	Monitor salinity levels, control pollution sources, and implement climate adaptation strategies
2	Water Quality	Pollution, industrial discharges, agricultural runoff	Enforce water quality standards, promote eco-friendly industrial practices, and adopt sustainable agriculture methods
3	Habitat Destruction	Urban development, salt production, infrastructure projects	Establish protected areas, enact land use planning and habitat restoration projects
4	Invasive Species	Introduction of non-native species, trade, human activities	Implement invasive species management plans and monitoring, and establish quarantine measures
5	Erosion Control	Deforestation, land degradation, human activities	Promote soil conservation practices, reforestation, and erosion control techniques
6	Wetland Vegetation	Destruction of vegetation, habitat disruption	Preserve and restore wetland vegetation, establish protected areas, and regulate human activities
7	Water Scarcity	Altered hydrology, increased demand for water	Promote water conservation, sustainable water use practices, and efficient irrigation methods
8	Climate Change Impact	Altered weather patterns, rising temperatures	Adaptation strategies, carbon reduction measures, and climate-resilient ecosystem management
9	Pollution Sources	Industrial discharge, agricultural runoff, urban sewage	Enforce pollution control regulations, promote responsible waste disposal and treatment measures
10	Altered Hydrology	Water diversion, infrastructure projects, changing hydrological patterns	Manage hydrological alterations, consider environmental impacts in development projects, and maintain natural flow regimes
11	Bird Disturbance	Human activities, tourism, habitat disruption	Implement bird-watching guidelines, regulate tourism activities, and establish bird sanctuaries
12	Overfishing	Excessive fishing, unsustainable practices	Implement and enforce fishing regulations, promote sustainable fishing methods, and maintain fish population monitoring
13	Sedimentation	Land use practices, construction activities	Control soil erosion, limit construction in sensitive areas, and improve sedimentation control techniques
14	Nutrient Imbalance	Agricultural runoff, wastewater discharge	Monitor nutrient levels, control pollution sources, and adopt nutrient management practices
15	Shoreline Erosion	Wave action, human activity, vegetation loss	Implement erosion control measures, establish buffer zones, and promote responsible land use

Table 9.1 (continued)

S. no.	Parameter of saline wetland	Causes of threats	Solutions for restoration
16	Microbial Balance	Pollution, altered hydrology, salinity changes	Monitor and maintain microbial communities, control pollution sources, and preserve natural flow regimes
17	Biodiversity Loss	Habitat destruction, pollution, invasive species	Establish protected areas, implement habitat restoration projects, and enforce pollution control regulations
18	Receding Water Levels	Climate change, water extraction, evaporation	Promote water conservation, manage water extraction, and adapt to changing climatic conditions
19	Avian Habitat Availability	Human disturbances, habitat loss, pollution	Preserve and restore avian habitats, enact regulations to limit disturbances, and control pollution sources
20	Altered Food Chains	Habitat loss, pollution, overfishing	Maintain natural food chains, control pollution sources, and promote sustainable fishing practices

cance. Inadequate support and investment in restoration efforts may result from many people failing to realise the biological value of these habitats or the benefits they give.

Limited Funding and Resources: Large-scale restoration efforts take extensive time, money, and skill. Saline wetland restoration can be difficult to fund, especially in areas with multiple demands on scarce resources. Restoration projects often struggle to get off the ground and stay viable if they lack sufficient funding.

Complex Ecological Interactions: Complex and impacted by many biological interactions, saline wetland habitats are fascinating to study. Understanding the interconnectedness of saltwater vegetation, soil chemistry, hydrological dynamics, and species interactions is crucial for the successful restoration of these ecosystems. Lacking this knowledge, restoration attempts might not have the expected ecological effects.

Invasive Species and Habitat Degradation: Invasive species pose a serious threat to the ecology of saline wetlands because they can outcompete local flora and fauna and disturb ecosystem processes. In addition, urbanisation, pollution, and overexploitation are all human-caused causes of habitat deterioration that can hinder restoration attempts. Successful restoration relies heavily on eradicating invasive species and repairing degraded habitats, but doing so can be difficult and time-consuming.

Climate Change and Sea-Level Rise: Restoration of saline wetlands is complicated by climate change impacts such as increasing sea levels, changing precipitation patterns, and more frequent and intense storms. Coastal erosion, flooding of salt marshes, and the demise of salt-tolerant plant life are all possible results of sea level rise. To preserve the long-term resilience of restored wetlands, restoration operations must take into account and adapt to these shifting environmental conditions.

Socioeconomic Factors and Land Use Conflicts: Human habitation and commercial pursuits including farming, fishing, and tourism frequently coexist alongside saline wetland ecosystems. Ecological goals of restoration might be difficult to reconcile with the desires of local people and stakeholders. Sustainable and inclusive restoration outcomes may be elusive in the face of potential conflicts over land use and resource allocation, calling for cautious negotiation and participatory approaches.

Long-Term Monitoring and Maintenance: Long-term monitoring and maintenance are essential for saline wetland restoration because they allow for the assessment of progress, strategy adjustments, and the maintenance of restored ecosystems. The success of restoration efforts, the identification of unintended repercussions, and the facilitation of adaptive management all depend on regular monitoring. However, long-term monitoring initiatives can be costly and call for consistent dedication and cooperation from all parties involved (Mitsch and Jørgensen 2003; Streever 1999; Telesetsky et al. 2016).

To overcome these obstacles, we need to employ a combination of stakeholder involvement, public education, robust finance, scientific inquiry, governmental backing, and flexible management techniques. It is essential that scientists, governments, local people, and conservation organisations work together to successfully restore and conserve saline wetlands in the face of these problems.

References

Allison SK (2023) Ecological restoration and environmental change: renewing damaged ecosystems. Taylor & Francis, Hoboken

Clewell AF, Aronson J (2012) Ecological restoration: principles, values, and structure of an emerging profession. Island Press, Washington, DC

Comín FA (ed) (2010) Ecological restoration: a global challenge. Cambridge University Press, Cambridge

Higgs E (2003) Nature by design: people, natural process, and ecological restoration. MIT Press, Cambridge

Holl K (2020) Primer of ecological restoration. Island Press, Washington, DC

Kent DM (ed) (2000) Applied wetlands science and technology. CRC Press, Boca Raton

Mitsch WJ, Jørgensen SE (2003) Ecological engineering and ecosystem restoration. John Wiley & Sons, New York

Streever B (1999) An international perspective on wetland rehabilitation. Springer Science & Business Media, Berlin

Telesetsky A, Cliquet A, Akhtar-Khavari A (2016) Ecological restoration in international environmental law. Taylor & Francis, Hoboken

UN Decade for Restoration

10

Abstract

The declaration of the "UN Decade on Ecosystem Restoration" marks a global endeavour to combat biodiversity loss and restore ecosystems, aligning closely with the goals of wetland restoration. This unit delves into practical methods for restoring saline wetlands, encompassing both traditional techniques like revegetation and innovative approaches to tackle salinity issues. It explores the transformative role of remote sensing technologies in monitoring and assessing saline wetland conditions, facilitating informed decision-making for restoration efforts. Additionally, the utilisation of Internet of Things (IoT) devices enables real-time data collection and analysis, offering insights into the dynamic nature of saline wetlands. Furthermore, artificial intelligence (AI) emerges as a powerful tool for data analysis, predictive modelling, and decision support in saline wetland restoration, optimising resource allocation for effective projects. Overall, this unit underscores the critical importance of wetland restoration, particularly in saline environments, and highlights the indispensable role of cutting-edge technologies in enhancing restoration endeavours, providing learners with essential knowledge and tools to contribute to this vital environmental cause.

Keywords

AI for restoration · IoT for restoration · UN for restoration · Wetland restoration

The United Nations has declared the 2020s to be the "Decade of Ecosystem Restoration", or the "UN Decade on Ecosystem Restoration", with the goal of speeding up the process of restoring degraded and devastated ecosystems worldwide. The UN General Assembly declared the start of the Decade on 1 March 2019, and it will last until 2030. It is a joint project of the UN Environment Programme (UNEP) and the UN Food and Agriculture Organization (FAO).

L. K. Sharma, R. Naik, *Conservation of Saline Wetland Ecosystems*,
https://doi.org/10.1007/978-981-97-5069-6_10

With the goal of reversing the damage humans have done to the planet's ecosystems, the United Nations has declared the 2020s the Decade of Restoration. Among the most important goals for the Decade are:

Catalysing Global Restoration Efforts: Forests, wetlands, meadows, mountains, and coastlines are just some of the habitats that the UN hopes to encourage and fund efforts to restore. Its goal is to galvanise support for expanding global efforts to restore ecosystems.

Raising Awareness and Building a Global Movement: Sustainable development, biodiversity conservation, climate change mitigation, and disaster risk reduction are only some of the goals of this Decade, which also aims to increase public understanding of the need to restore ecosystems. Its ultimate goal is to launch a worldwide restoration movement that includes government, civil society, indigenous communities, young people, and the commercial sector.

Facilitating Knowledge Exchange and Capacity Building: It intends to make it easier for people to share what they've learned about restoring ecosystems. It encourages the creation and distribution of scientifically sound data and tools to back up restoration initiatives. It is also planned to implement capacity-building programmes to increase the knowledge and abilities of practitioners engaged in restoration work.

Monitoring and Reporting on Restoration Progress: The goal of this Decade is to set up a system to keep tabs on global restoration activities. Through the creation of indicators and reporting approaches, progress may be assessed and new difficulties can be found in the struggle to restore ecosystems.

Mobilizing Funding and Investment: The goal of this Decade is to raise awareness and funds for restoring ecosystems. To that end, it promotes collaboration and novel means of funding for environmental revitalisation efforts.

There are several international agreements that coincide with the United Nations Decade on Ecosystem Restoration. These include the Bonn Challenge, the New York Declaration on Forests, and the Sustainable Development Goals (SDGs). To counteract biodiversity loss, and climate change, and to achieve sustainable development, it is imperative that ecosystems be restored immediately.

The United Nations Decade on Ecosystem Restoration aspires to promote transformative action, collaborative partnerships, and policy reforms that can aid in the recovery and resilience of ecosystems worldwide by making their restoration a top priority on the international agenda (Craft 2022; Finlayson et al. 2020).

10.1 Technologies for Restoration

10.1.1 On-Field Methods of Saline Wetland Restoration

Multiple on-site techniques are used to restore a saline marsh, depending on the site's demands and characteristics. Common field techniques for restoring salt marshes are given in Fig. 10.1 and described in the following sections:

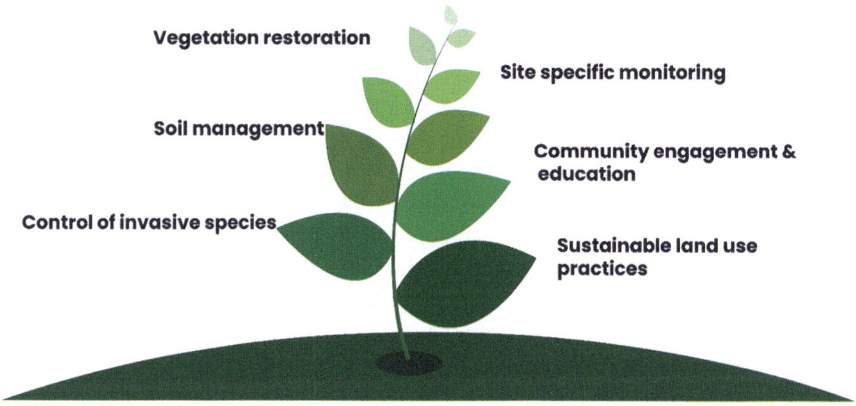

Fig. 10.1 On-field methods of saline wetland restoration

Saline wetland restoration projects frequently centre on re-creating natural hydrological conditions. In order to restore the natural water levels, salinity gradients, and water exchange patterns within the wetland, it may be necessary to alter or reconnect water channels, ditches, or tidal flows. Levees, dams, and culverts can all play a role in this, either by being built or dismantled.

Vegetation Restoration: For saline wetlands to continue operating normally and without collapsing, salt-tolerant plants must be reintroduced. Planting native salt-tolerant plants, restoring or fostering the expansion of existing seed banks, and managing invasive plant species are all ways to achieve this goal. Soil stabilisation, increased biodiversity, wildlife habitat, and improved nutrient cycling are just a few of the many benefits of restoring vegetation.

Soil Management: The restoration of saline wetland soils is complicated by their unique properties. In order to enhance the soil's structure, fertility, and water-holding ability, it may be necessary to use soil supplements like gypsum or organic matter as part of a comprehensive approach to soil management. Erosion control blankets and erosion-resistant plant plantings are two methods that can be used to prevent soil erosion.

Control of Invasive Species: Outcompeting native vegetation and disrupting ecosystem processes, invasive plant species pose a threat to the viability of salty wetland restoration. Invasive species can be managed through harvesting by hand, spraying with herbicides, or using biological controls. This aids in halting the spread of invasives and making way for the growth and development of natives.

Monitoring and Adaptive Management: Maintaining a saline wetland after restoration requires constant observation and flexible administration. Water quality, vegetation composition, and wildlife populations are just a few examples of crucial indicators that can be monitored on a regular basis to gauge the success of restoration activities. To maximise results and solve unforeseen problems, adaptive management entails modifying restoration procedures depending on monitoring data.

Community Engagement and Education: For saline wetland restoration to be effective and long-lasting, it is crucial to involve local communities and other stakeholders. Including locals, such as indigenous peoples and traditional land users, in the restoration planning process is essential for incorporating local knowledge and values. Stewardship and support for continuing restoration efforts can be cultivated through community education and awareness campaigns.

Sustainable Land Use Practices: To slow the rate of degradation and aid in the restoration process, it is essential to encourage sustainable land use in the immediate vicinity of the wetland. Sustainable farming methods, encouraging eco-friendly vacations, and careful waste management and land-use planning are all examples of measures that can reduce environmental impact.

When restoring a salty wetland, the approach taken will be determined by the type of wetland, its location, and the biological goals that must be accomplished. When restoring a wetland, it's important to work with specialists, stakeholders, and local people to find the best approach based on the wetland's unique needs and goals (Goreau and Trench 2012; Upadhyay et al. 2019).

10.1.2 Remote Sensing Use for Saline Wetland Restoration

Saline wetland restoration can benefit greatly from the use of remote sensing because of the information it provides on the wetland's current condition, its ability to track changes over time, and its ability to direct restoration efforts. Some examples of how remote sensing is applied to rehabilitating saline wetlands are given in Fig. 10.2 and some examples in Table 10.1 and detailed description are as follows:

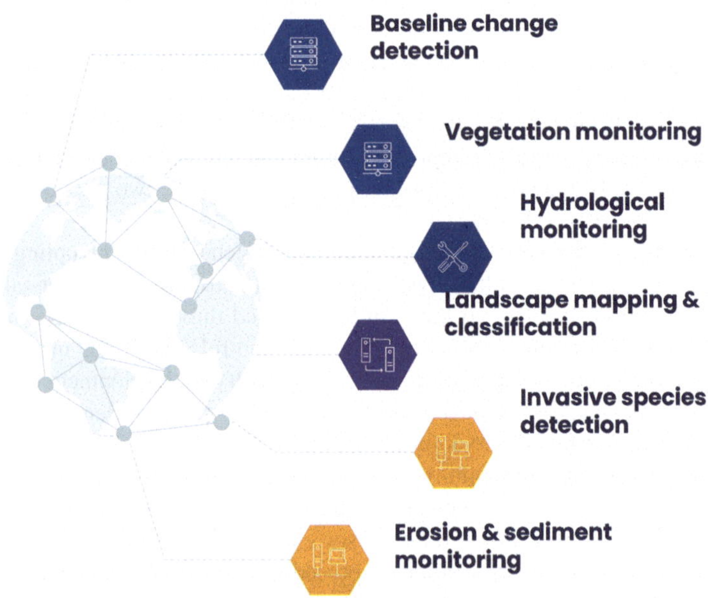

Fig. 10.2 Use of remote sensing for saline wetlands restoration

Table 10.1 Role of remote sensing for wetland restoration

S. no.	Parameter detectable through remote sensing	Method of detection	Use for saline wetland restoration
1	Salinity levels	Spectral reflectance analysis	Monitor and manage salinity for habitat restoration
2	Vegetation health	Normalised difference vegetation index (NDVI)	Assess vegetation health for restoration planning
3	Water quality	Water absorption bands analysis	Monitor water quality and identify pollution sources
4	Soil moisture content	Thermal infrared and microwave sensing	Control soil moisture for vegetation restoration
5	Wetland extent	Multispectral and radar imaging	Assess wetland boundaries for restoration planning
6	Habitat mapping	Hyperspectral imaging and GIS	Identify critical habitats for conservation efforts
7	Invasive species detection	Hyperspectral and multispectral imaging	Detect non-native species for removal and restoration
8	Erosion mapping	LiDAR (light detection and ranging)	Identify areas prone to erosion for restoration
9	Water depth	Radar and LiDAR technology	Monitor water depth and fluctuations for habitat planning
10	Microtopography	LiDAR and aerial photogrammetry	Restore natural microtopography for wetland health
11	Sedimentation levels	Remote sensing and GIS	Monitor sediment deposition for erosion control
12	Bird migration patterns	Radar and GPS tracking	Understand bird movements for habitat conservation
13	Algal blooms	Multispectral imaging and water quality analysis	Detect and mitigate harmful algal blooms
14	Soil salinity	Hyperspectral imaging	Identify saline areas for targeted restoration
15	Water temperature	Thermal infrared imagery	Monitor temperature for ecological health
16	Nutrient levels	Water quality and spectral analysis	Detect nutrient imbalances for restoration planning
17	Land use changes	Satellite imagery and land cover classification	Assess human impacts on wetlands for restoration
18	Hydrological patterns	Satellite and radar data analysis	Understand flow patterns for wetland management
19	Shoreline changes	Aerial photography and GIS	Monitor shoreline erosion and plan restoration measures
20	Microbial community	DNA sequencing and hyperspectral data	Assess microbial diversity for ecosystem health

Wetland Mapping and Classification: Saline wetland areas can be mapped and categorised with the help of remote sensing imaging like aerial photography and satellite images. These photos show the whole scope of the region's wetlands, including their borders and the many wetland kinds present there. The distribution of wetlands can be mapped in order to better plan and prioritise restoration efforts and locate problem regions.

Vegetation Monitoring: The vitality and distribution of vegetation in salt marshes can be evaluated by means of remote sensing. Changes in vegetation density, composition, and distribution can be tracked by examining spectral signatures, vegetation indices, and vegetation cover estimations generated from remote sensing data. This data is helpful for determining which regions have been restored successfully and which still need attention.

Hydrological Monitoring: Hydrological properties of salt marshes can be gleaned from remote sensing data collected by radar or optical images. This involves keeping an eye on water levels, spotting variations in water coverage, and assessing salinity and other water quality parameters. So as to restore natural water regimes and promise proper salinity gradients within the wetland, remote sensing data can aid in identifying places that need hydrological interferences.

Erosion and Sedimentation Monitoring: Monitoring outlines of erosion and alluviation in salt marshes is likely with the aid of remote sensing. Shoreline shifts, zones of high erosion, and sediment deposition can all be traced by comparing old and new snapshots. During the restoration phase, this data is valuable for making decisions about sediment management and building erosion control procedures.

Invasive Species Detection and Monitoring: Invasive species in saline wetlands can be found and tracked with the support of remote sensing technology. The occurrence of invasive species can be distinguished and their spread assessed through the use of remote sensing data by inspecting spectral properties and vegetation indicators. Invasive species can have a harmful effect on wetland ecosystems, but early diagnosis allows for tailored management activities to suppress them.

Baseline and Change Detection Analysis: Before attempting to restore saltwater wetland ecosystems, remote sensing imagery can be used to determine the current state of the ecosystem. Wetland extent, vegetation cover, water quality, and other pertinent factors can all be evaluated in relation to restoration efforts with the use of this baseline data by comparing photos taken before and after restoration.

Planning and Decision Support: Planning and decision-making in salty wetland restoration are aided by remote sensing and geospatial technology. When combined with supplementary spatial data, such as land cover, terrain, and infrastructure, the information can be used to evaluate restoration choices, locate suitable restoration locations, and measure the potential influences and benefits of restoration efforts.

The use of remote sensing technology has made it possible to collect data from previously inaccessible or extremely broad areas at a reasonable cost. It provides a broader and continuous perspective of the wetland environment, supplementing field surveys and ground-based monitoring. Understanding saline wetland dynamics, making decisions based on facts, and putting restoration strategies into action are all improved by combining remote sensing with other ecological and socio-economic data.

10.1.3 Use of IoT for Saline Wetland Restoration

By allowing for real-time monitoring, data collecting, and management of many environmental conditions, the IoT can play a crucial role in saline wetland restoration of which some examples are given in Fig. 10.3 and some examples in Table 10.2. Possible Internet of Things uses in rehabilitating saline wetlands include as follows:

Water Quality Monitoring: Salinity, pH, dissolved oxygen, and nutrient concentrations are just some of the water quality metrics that can be tracked by Internet of Things (IoT) sensors in salt marshes. At predetermined intervals, these sensors can collect data and send it wirelessly to a control hub. When water quality problems arise, their causes may be pinpointed and their effects can be mitigated more quickly thanks to real-time monitoring.

Hydrological Monitoring: Water levels, flow rates, and tide movements are only some of the hydrological characteristics that may be tracked with the help of Internet of Things sensors in salt marshes. In order to evaluate the efficacy of hydrological restoration efforts and to guarantee the maintenance of suitable water regimes inside the wetland, sensors can be deployed to provide continuous data on water levels and fluctuations. Water levels, flow rates, and tide movements are only some of the hydrological characteristics that may be tracked with the help of Internet of Things sensors in salt marshes. In order to evaluate the efficacy of hydrological restoration efforts and to guarantee the maintenance of suitable water regimes inside the wetland, sensors can be deployed to provide continuous data on water levels and fluctuations.

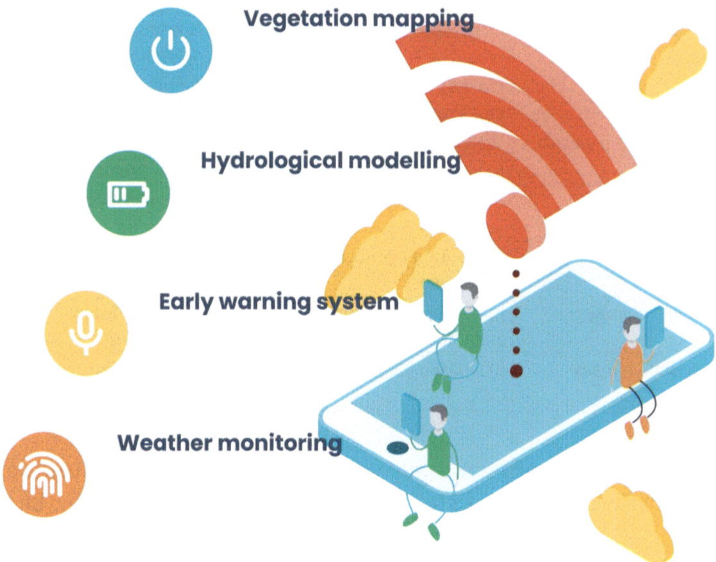

Fig. 10.3 Use of IoT for saline wetlands restoration

Table 10.2 Use of IoT for saline wetland restoration

S. no.	Parameter detectable through IoT	Method of detection	Use for saline wetland restoration
1	Water level	IoT-connected sensors	Monitor water levels for habitat and hydrology management
2	Salinity levels	Salinity probes and sensors	Control and adjust salinity for habitat restoration
3	Water quality	IoT-based water quality sensors	Detect pollution sources and maintain water quality.
4	Soil moisture content	Soil moisture sensors	Manage soil moisture for vegetation restoration
5	Vegetation health	IoT-based cameras and vegetation indices	Assess vegetation health and plan for restoration
6	Wetland extent	Remote sensing and GPS	Monitor wetland boundaries for restoration planning
7	Erosion mapping	IoT-based erosion sensors	Identify erosion-prone areas for restoration measures
8	Microclimate conditions	Weather stations and IoT sensors	Understand microclimate variations for restoration planning
9	Water temperature	Temperature sensors	Monitor temperature variations for ecosystem health
10	Nutrient levels	IoT-based water quality sensors	Detect nutrient imbalances for restoration planning
11	Algal blooms	Algal bloom sensors and water quality IoT	Detect and mitigate harmful algal blooms
12	Bird migration patterns	IoT-based tracking and monitoring systems	Understand bird movements for habitat conservation
13	Sedimentation levels	IoT-based sediment sensors	Monitor sediment deposition for erosion control
14	Microbial community	DNA sequencing and IoT sensors	Assess microbial diversity for ecosystem health
15	Invasive species detection	IoT-based cameras and machine learning	Detect non-native species for removal and restoration
16	Hydrological patterns	Water flow sensors and IoT data analysis	Understand hydrological patterns for wetland management.
17	Water depth	IoT-based depth sensors	Monitor water depth and fluctuations for habitat planning
18	Soil salinity	Soil salinity probes and IoT sensors	Identify saline areas for targeted restoration
19	Land use changes	IoT-connected cameras and GIS	Assess human impacts on wetlands for restoration planning
20	Shoreline changes	IoT-based cameras and GIS	Monitor shoreline erosion and plan restoration measures

Weather Monitoring: Temperature, humidity, wind speed, and precipitation are only some of the meteorological characteristics that can be tracked by Internet of Things weather stations set up in salt marshes. Understanding the wetland's microclimate, spotting extreme weather occurrences, and quantifying their impacts are all made easier with access to real-time meteorological data. Adaptive management tactics and the prediction of prospective restoration difficulties benefit greatly from this data.

Vegetation Monitoring: Vegetation in salt marshes can be monitored with the help of IoT devices, such as remotely piloted aerial vehicles. High-resolution images and spectral data collected by these sensors can be used to evaluate plant health, distribution, and seasonal changes. Such information is useful for gauging the effectiveness of plant life restoration initiatives, identifying invasive species, and directing precise treatments.

Wildlife Monitoring: Saline wetland wildlife populations and activity levels can be tracked with the help of IoT devices like remote cameras and acoustic sensors. All sorts of information about the presence, behaviour, and abundance of species can be gathered with the use of these gadgets. Conservation efforts can be better informed by learning more about wildlife and their habitat preferences through monitoring programmes.

Soil Monitoring: Soil moisture, temperature, and nutrient levels may all be monitored with the help of IoT sensors. In order to evaluate soil conditions, optimise irrigation methods, and guarantee suitable soil conditions for plant growth, continuous soil monitoring is essential. The restoration of vegetation and the overall health of wetlands depend on this data.

Early Warning Systems: Potential dangers to salt marshes can be identified and dealt with using Internet of Things-based early warning systems. Water level sensors, for instance, can sound the alarm if the water level suddenly rises or falls, signalling the potential for a breach or floods. With the help of early warning systems, we can take preventative measures in a timely manner, reducing the potential for damage to wetland ecosystems.

Data Integration and Analysis: Using advanced analytics and machine learning methods to combine the massive amounts of data produced by IoT devices with other environmental and spatial datasets can yield useful insights. Saline wetland restoration efforts can benefit from data analysis by revealing trends, correlations, and anomalies that can then be used to make educated decisions and employ adaptive management strategies.

IoT has the ability to automate saline wetland restoration through real-time monitoring, data-driven decision-making, and other means. It can aid in the restoration and long-term sustainability of saline wetlands by providing continuous and precise data on many environmental parameters that are essential to their management and conservation.

10.1.4 Use of AI for Saline Wetland Restoration

Multiple applications exist for the use of AI in salty wetland restoration. Probable uses of AI in naturalising saline wetlands are given in Fig. 10.4 and some examples in Table 10.3:

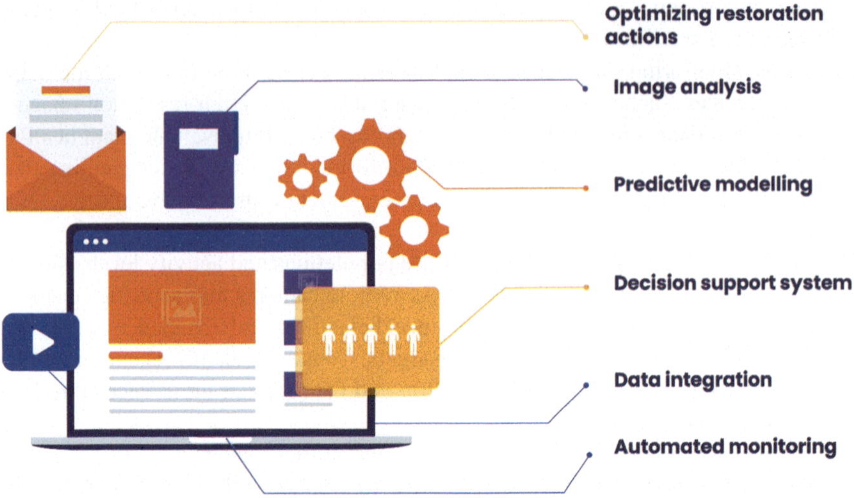

Fig. 10.4 Use of AI for saline wetlands restoration

Image Analysis and Classification: Land cover types in salt marshes can be classified using AI algorithms trained on satellite datasets, aerial photographs, or drone outputs. This can aid researchers in mapping wetlands, classifying plants, and tracking variations in wetland vegetation. AI algorithms can quickly and precisely classify land cover by processing enormous amounts of data.

Species Detection and Monitoring: Species-specific detection and monitoring in saline wetlands is possible with the help of computers and machine learning. Photographs or acoustic recordings can be used to train AI algorithms to identify and track the occurrence of threatened or invasive species. This is useful for keeping track of population changes, learning about how diverse species interact, and developing more detailed management plans.

Predictive Modelling: Predictive models of how saline wetlands will respond to multiple conditions can be created using AI algorithms. Forthcoming wetland conditions and the impacts of restoration efforts can be predicted by using AI models to examine past data on hydrological forms, vegetation dynamics, and climate parameters. This is important for selecting choices, organising restoration plans, and gauging the success of restoration actions.

Data Integration and Analysis: AI can combine and investigate many types of environmental data to spot drifts, correlations, and forms in saline wetlands. Remote sensing, meteorological data, water quality sensors, and other sources of information can be combined by AI to provide light on the intricate interplay of elements influencing wetland status. Restoration inventiveness and adaptive management practices can both profit from this data.

Optimization of Restoration Actions: It is possible to amplify the effectiveness of restoration attempts in saline wetlands by employing AI optimisation algorithms. These algorithms can allow for a wide variety of factors—from financial restrictions and ecological objectives to site-specific situations—to determine the best course of action for restoration. With the help of AI optimisation, restoration

Table 10.3 Use of AI for saline wetland restoration

S. no.	Parameter detectable through AI	Method of detection	Use for saline wetland restoration
1	Salinity levels	Machine learning from spectral data	Monitor and manage salinity for habitat restoration
2	Vegetation health	Image analysis and machine learning	Assess vegetation health for restoration planning
3	Wetland extent	Satellite imagery and object detection	Assess wetland boundaries for restoration planning
4	Invasive species detection	Image recognition and machine learning	Detect non-native species for removal and restoration
5	Erosion mapping	Aerial and satellite imagery analysis	Identify areas prone to erosion for restoration
6	Water quality	AI analysis of water quality data	Monitor water quality and identify pollution sources
7	Soil moisture content	AI-based soil moisture modelling	Control soil moisture for vegetation restoration
8	Bird migration patterns	AI-based tracking and monitoring systems	Understand bird movements for habitat conservation
9	Algal blooms	Image analysis and machine learning	Detect and mitigate harmful algal blooms
10	Water temperature	AI analysis of temperature data	Monitor temperature variations for ecosystem health
11	Nutrient levels	AI analysis of water quality data	Detect nutrient imbalances for restoration planning
12	Microbial community	AI analysis of DNA sequencing data	Assess microbial diversity for ecosystem health
13	Hydrological patterns	AI-based analysis of hydrological data	Understand flow patterns for wetland management
14	Water depth	AI-based analysis of depth data	Monitor water depth and fluctuations for habitat planning
15	Sedimentation levels	AI analysis of sediment data	Monitor sediment deposition for erosion control
16	Infrared imaging	AI-based thermal imaging analysis	Detect changes in microclimate conditions for restoration planning.
17	Inundation mapping	AI-based flood mapping and prediction	Manage water inundation for habitat conservation
18	Microplastic pollution	AI image analysis for microplastic identification	Detect and mitigate microplastic pollution
19	Shoreline changes	AI analysis of shoreline imagery	Monitor shoreline erosion and plan restoration measures
20	Vegetation density	AI-based analysis of vegetation density	Monitor and manage vegetation for habitat restoration
21	Water flow patterns	AI-based analysis of flow data	Understand water flow patterns for wetland management

activities may be prioritised, resources can be assigned, and desirable ecological results can be realised.

Decision Support Systems: Restoration professionals can benefit from the development of AI-based decision assistance tools. Recommendations on appropriate restoration procedures, timing of interventions, potential dangers, and trade-offs can be generated by these systems thanks to the incorporation of diverse data sources, models, and expert knowledge. Artificial intelligence decision support tools improve the effectiveness of restoration planning and management.

Automated Monitoring and Maintenance: By combining IoT sensors, remote sensing tools, and machine learning algorithms, AI enables automated monitoring of salt marshes. AI systems can monitor sensor data in real time, look for outliers, and respond with warnings or predetermined actions when those thresholds are surpassed. Wetland restoration management can be more efficient and effective if changes in the environment, threats, or maintenance needs are identified on time (Van Andel and Aronson 2012; Van Dover et al. 2014).

References

Craft C (2022) Creating and restoring wetlands: from theory to practice. Elsevier, Amsterdam

Finlayson CM, Rastogi G, Mishra DR, Pattnaik AK (eds) (2020) Ecology, conservation, and restoration of Chilika Lagoon. Springer, New Delhi

Goreau TJ, Trench RK (eds) (2012) Innovative methods of marine ecosystem restoration. CRC Press, Boca Raton

Upadhyay AK, Singh R, Singh DP (eds) (2019) Restoration of wetland ecosystem: a trajectory towards a sustainable environment. Springer, Berlin

Van Andel J, Aronson J (eds) (2012) Restoration ecology: the new frontier. John Wiley & Sons, New York

Van Dover CL, Aronson J, Pendleton L, Smith S, Arnaud-Haond S, Moreno-Mateos D et al (2014) Ecological restoration in the deep sea: desiderata. Mar Policy 44:98–106

The manufacturer's authorised representative in the EU is Springer Nature Customer Service Centre GmbH, Europaplatz 3, 69115 Heidelberg, Germany. If you have any concerns regarding our products, please contact ProductSafety@springernature.com

Printed and bound by CPI Group (UK) Ltd, Croydon, CR0 4YY

01/10/2026

02225857-0005